Practical Radio Resource Management in Wireless Systems

The Artech House Universal Personal Communications Series

Ramjee Prasad, Series Editor

For a listing of recent titles in the *Artech House Mobile Communications Library*, turn to the back of this book.

Practical Radio Resource Management in Wireless Systems

Sofoklis A. Kyriazakos
George T. Karetsos

Artech House, Inc.
Boston • London
www.artechhouse.com

Library of Congress Cataloging-in-Publication Data
Kyriazakos, Sofoklis A.
    Practical radio resources management in wireless systems/Sofoklis A. Kyriazakos.
    p. cm.—(Universal personal communications)
    Includes bibliographical references.
    ISBN 1-58053-632-8 (alk. paper)
    1. Telecommunication—Traffic—Management. 2. Personal communication service
systems. 3. Radio frequency allocation—Planning. 4. Mobile communications systems.
5. Wireless communication systems. I. Title. II. Artech House universal personal
communications series.

TK5102.985.K97   2004
384.5'34'068—dc22   2004041023

British Library Cataloguing in Publication Data
Kyriazakos, Sofoklis A.
    Practical radio resource management in wireless systems.— (Artech House universal
personal communications library)
    1. Wireless communication systems
    I. Title   II. Karetsos, George T.
    621.3'8456

ISBN 1-58053-632-8

Cover design by Yekaterina Ratner

© 2004 ARTECH HOUSE, INC.
685 Canton Street
Norwood, MA 02062

International Standard Book Number: 1-58053-632-8

10 9 8 7 6 5 4 3 2 1

Contents

Preface

Radio resource management (RRM) in wireless telecommunication systems is of prime importance for current and future mobile network operators. Efficient resource management is strongly related to increased network performance, which is, in turn, linked with customer satisfaction and operator revenue.

In the early 1990s, when second generation (2G) cellular systems were introduced, the global system for mobile communications (GSM) was criticized as over-specified, since a large number of its features were thought to be redundant for the limited number of subscribers at the time. After more than 10 years of operation, it is now clear that almost all of the offered GSM capabilities are heavily utilized. Furthermore, with the introduction of new services many more features are being requested—along with more bandwidth—to accommodate efficiently all the emerging needs. The above developments paved the way for the introduction of the third generation (3G) of wireless telecommunication systems whose main characteristics are an increased capacity in the wireless links and an emphasis on the provision of efficient data communications. At the same time research is already under way for the definition and specification of the fourth generation (4G) of wireless telecommunications, although the primary characteristics of this generation have yet to be determined. Mobile communication networks are continuously expanding in size as traffic volumes and supported services also increase. However, the rapid evolution of wireless technologies does not guarantee that the respective systems will perform in the way their designers intended. This problem has been experienced repeatedly in the past with shortcomings appearing often and unexpectedly.

Congestion and its consequences on wireless communication networks is one of the hottest issues—of nightmarish proportions—for mobile network operators. Despite the efforts to deal with this problem, undertaken so far both by the industry and the research community, congestion continues to appear especially in critical situations like floods, earthquakes, or accidents, when communication is most needed.

Driven by these facts, in 1999 we initiated a set of research activities and studies that received a warm welcome from the cellular wireless networking community. This book details the results of these four years of work. We present and analyze how resource management mechanisms can increase the efficiency of cellular networks, especially during periods of overload. The techniques presented here are practical in the sense that most of them can easily be applied on operational networks. The resource management techniques are compared by focusing on key performance indicators (KPIs), such as call blocking probability, call set up success rate, and other relevant measures that can be monitored directly by cellular operators and indirectly by the subscribers, defining user satisfaction. Additionally, we present system architectures that are capable of effectively accommodating and supporting these congestion management techniques for present and future-generation wireless telecommunication systems.

The book is organized as follows. Chapter 1 introduces resource management and analyzes the importance of RRM. The arguments are presented both from the side of cellular network operators as well as from the side of mobile users. A brief description of possible network shortcomings is presented in preparation for the in-depth analysis of the subsequent chapters of the book. Chapter 2 analyzes RRM for 2G wireless systems. First, we give an architectural description of the envisaged system, which enhances GSM with dynamic resource management capabilities. Then, we provide an extensive statistical evaluation based on real measurements performed to develop a network management platform that represents well-configured cellular systems. The statistical evaluation highlights the limitations of GSM systems, and characteristic traffic load scenarios are extracted and modeled. Subsequently, Chapter 2 presents RRM techniques and implementation guidelines for cellular operators. The resource management techniques are validated through measurements on a real GSM network even when the nature of the technique does not allow for the use of simulators.

Chapter 3 presents RRM for 2.5 generation (2.5G) wireless systems. At the beginning we introduce this technology followed by an evaluation its network performance. This evaluation focuses on characteristic diagrams that point to the possible network shortcomings that we would like to tackle. Since traffic volume is expected to become worse due to the increase in data services, innovative resource management solutions are presented for 2.5G.

Chapter 4 deals with the congestion problem of 3G wireless systems, first presenting an architectural overview of universal mobile telecommunications system (UMTS). Since no previous experience in UMTS exists, statistical evaluation is not possible. Therefore, only extensive studies regarding network congestion and possible associated shortcomings will be presented. Since RRM in UMTS is a key topic, techniques and implementation guidelines will be presented.

Chapter 5 introduces wireless systems beyond 3G as well as prospective resource management techniques. The definition of a beyond-3G system is still not clear, although several proposals have been discussed for candidate 4G networks. The most possible roadmap is the seamless interoperability between heterogeneous wireless networks as the transition of UMTS to 4G. Thus, we focus on diversified radio environments consisting of cellular networks, wireless local area networks (WLANs) and broadcasting systems for fixed wireless access as candidates for integration into a wide 4G system. RRM is also very important in an environment where different networks have to cooperate and adapt their capabilities efficiently. Accordingly, we also discuss a set of scalable architectures.

Chapter 6 focuses on the business models that can be applied for the implementations presented in the previous chapters. To develop these models, we present KPIs for wireless systems as well as the results of their evaluation. Particularly in composite radio environments of the 4G, business models are expected to play a significant role, first for the selection of the suitable network segments and subsequently for resource management.

Our intention here is not only to analyze resource management from a technical point of view, but also to present the business opportunities that may arise when such provisions exist. The final outcome is, we hope, a practical book that can be used by researchers, network operators, and high-level management people in the area of wireless telecommunications.

Finally, we would like to acknowledge the CAUTION and CAUTION++ projects, which were partially funded by the European Commission. Some of the studies presented in this book were performed under the framework of the research and development activities of these projects. In addition we would like to acknowledge the following colleagues for contributions to individual sections:

- Evangelos Gkroustiotis (Chapters 2–5);
- Kostas Vlahodmitropoulos (Chapters 2–4);
- Dimitris Nikitopoulos and Nikos Papaoulakis (Chapters 2 and 3);
- Charis Kechagias (Chapter 2);
- Sami Nousiainen and Krzystof Kordybach (Chapter 4);
- Antonis Markopoulos (Chapter 5);
- Ivan Mura (Chapter 6).

We would also like to thank the publisher of the book and everyone who supported our work.

CHAPTER 1

Why Resource Management?

In recent years, we have witnessed the "wireless revolution." This has engendered an unimaginable increase in the number of mobile subscribers, which sometimes even surpasses the number of users of fixed networks. At the same time, the services available to mobile users have evolved rapidly, and it would appear that multimedia communication while on the move is not such a distant goal. However, these developments require new prerequisites from the underlying networking infrastructure where radio resources are scarce and already overexploited and the existing architectures lack the flexibility that would allow the easy and efficient introduction of emerging services.

The inherent characteristics of wireless networks make the maintenance of a consistent network performance one of the most important issues for the respective operators, whose main concern is to satisfy their subscribers by providing the quality of service (QoS) requested. To adequately provide for their networks, operators usually estimate future needs based, principally, on the increase in the number of subscribers and, subsequently, on restructuring the system by adding more resources where needed. Thus, it is apparent that the planning activity of a mobile network operator is a continuous process and has been performed until recently by focusing on the long-term needs of network users. However, under this approach operators are unable to accommodate a sudden and transient increase in traffic. Traffic overload arises every day, and, to some extent, it is predictable; therefore, the operator knows that the demand for communication increases during rush hours in specific places. On the other hand, there are several unpredictable traffic load scenarios such as accidents, strikes, demonstrations, and catastrophes when the demand for communication resources is considerably higher than it is during normal operation. In this case the inability of the operators to serve their subscribers has been strongly criticized. Whenever operators wish to increase the maximum accommodated traffic, they must consider the optimum trade-off between equipment and installation costs, as well as depreciation period and user satisfaction. Of these parameters, the latter one is the most important and least easily quantified with respect to call minutes or revenues.

In the following chapters, extensive studies, based on real measurements, will clearly show that the maximum traffic that can be accommodated is not a constant value, since it varies according to the traffic load scenario. It appears that in most of the congestion scenarios, radio networks remain underutilized but do not allow subscribers to set up calls, while, at the same time, the operators are unable to serve a large number of requests, despite the fact that network resources are still available.

The dimensioning of these networks is usually based on rough estimates rather than accurate models, and this can lead to inefficient system configuration and deployment [1]. These facts prove that planning based on a simplistic provisioning of resources of predictions of future demands has failed to effectively meet the expectations of the subscribers—although they ensure that networks remain operational.

Until now we have seen that the efforts for improving the QoS in cellular networks have focused mainly on the physical resources whereas the efficient management, dimensioning, and usage of the logical resources available within these systems are rarely considered. In our work we focus on how to manage efficiently the resources of operational cellular networks by enhancing and fine-tuning the already installed network management system (NMS). Additionally, we propose innovative mechanisms, promising enhanced capacity management, especially in critical situations.

There are three ways for operators to enhance their system and increase network performance. The first one is to improve radio planning, either by adding new base stations, or by modifying the parameters and transmitters in the existing sites. The second solution to achieve increased network performance is the static reconfiguration of the network parameters. Finally, the third way is for operators to introduce adaptive resource management techniques, supported by enhanced monitoring of the network and intelligent decision making; this would enable a dynamic system reconfiguration. In the following sections of this chapter, these mechanisms are further analyzed.

1.1   Enhancing Radio Planning

Enhancing radio planning is the most common mechanism to increase the system's performance. The operator monitors the system performance constantly and stores the data in the network data warehouse (DW). The data can be postprocessed and shows on a daily, weekly, or monthly basis the behavior of the KPIs of the network. In case KPIs fall below predefined satisfaction thresholds in a physical area, network resources can be reallocated in a way that supports traffic demand. This can be achieved in systems based on frequency division multiple access (FDMA), such as GSM, by adding transceivers (TRXs) in the existing sites, thereby offering additional communication channels. The procedure requires that additional pairs of frequencies in the same area coexist without causing any interference. In the event that this cannot be avoided, the operator can even change the frequency allocation of the adjacent cells. If the problem is not only an increased demand for traffic, but also insufficient coverage, then additional base stations might need to be placed, or existing ones should be modified. This can also be done to increase the maximum traffic that can be accommodated, requiring additional equipment and effort. The completion of this procedure requires monitoring of the area to detect whether network performance is increased and to check that it does not cause any shortcomings to the neighbor cells.

To summarize, this procedure is costly and the depreciation period can even be a full calendar year, while a small alteration in the area (i.e., the appearance of new shops and offices) can result in a new demand from the planning activity.

1.2   Enhancing Network Planning

Radio planning is one of the main means available to the cellular network operator to enhance its system's performance and to confront network shortcomings. The adjustment of network parameters, which should sometimes be performed before radio planning, is another process that improves the network's performance. Cellular systems support a large number of mechanisms that allow the success of wireless digital communications. Handover, power control, reselection hysterisis, and admission control are four typical procedures in wireless systems, most recently integrated into the UMTS architecture's protocols. Proper configuration of these parameters is essential. For example, the ping-pong effect often observed in cellular systems during handovers can lead to an increase in call dropping and congestion and low communication quality.

Therefore, network planning and optimization, without any additional installation of equipment, can be a low-cost solution for network improvement. As a result, reconfiguring the network parameters cannot solely guarantee increased performance, as the network does not always respond positively to all traffic overload situations.

1.3   Adaptively Managing the System

The approach for optimizing the system is based mainly on the preceding process (described in Section 1.2), and it is not related to radio planning. It requires a high-performance, real-time network monitoring system that observes a set of KPIs and enables intelligent decision-making with respect to the network's current state. It must be able to detect all possible traffic overload situations and determines how network parameters can be reconfigured spontaneously as needed. In the GSM system, mobile terminals in idle mode continuously listen to broadcast messages that provide information that can be used to select the appropriate cell for a call set up. By changing these broadcast messages, a mobile terminal can consider the cell to be larger or smaller. If the cell appears to be smaller, the terminal will actually transmit with the same power as before, and therefore, fewer users will attempt to set up a call or attach to the cell; this effectively reduces the possibility of congestion. Therefore, the classification of congestion situations according to the size, duration, and impact on radio resources (as described in the following chapters) shows that different scenarios require different cell sizes, when the *cell-breathing* resource management technique is applied.

1.4   Conclusions

This chapter briefly presented the reasons why resource management is important for wireless systems for the present and future generations. In addition, we discussed the three principal methods of enhancing the performance of wireless systems. Cellular operators are currently using the first two methods to achieve improved network performance. Our approach is more practical and adaptive, since it decreases costs by not requiring continuous network planning, and it is based on a self-trained intelligent system. As a result, this method reconfigures network parameters and improves the stability of the network. This scalable approach is an evolution in system's dynamic reconfiguration. In the following chapters we will present resource management techniques as well as their implementation in a novel system that supports both monitoring and decision-making. It should be mentioned that, despite this evolutionary approach, each resource management technique is also independent from this system and can be applied manually. This, in fact, is the procedure currently followed by network operators.

Reference

[1]   Uskola, J., *Strategic Network Planning Process in Transition from GSM to UMTS Technology*, Master's Thesis, Helsinki University of Technology, Espoo, Finland, 2000.

CHAPTER 2

Resource Management in 2G Wireless Systems

This chapter deals with the issue of radio resource management in 2G wireless systems. Section 2.1 briefly presents the structure of a 2G network, focusing on the air interface, since the management techniques aim to solve the traffic congestion problem in the wireless part of the network. Furthermore, we present an in-depth performance evaluation study of a GSM system, based on measurements performed in a real networking environment. This is followed by a classification of congestion situations in traffic load scenarios, so that any time this occurs, we follow a predefined procedure to solve it. Finally, a number of resource management techniques are presented and analyzed.

2.1 Architecture of 2G Systems

The GSM system as we know it today, started as an effort to develop a standardized Pan-European mobile radio network that would replace the various incompatible systems that were already under operation in the early 1980s in many European countries. This important decision was taken in 1982 during a meeting of the Conférence Européene des Postes et des Télécommunications (CEPT) which formed a study group called the Groupe Spécial Mobile (GSM[1]). It was also agreed that the new system should be compatible with the wireline networks [*integrated services digital network* (ISDN) and *International Standards Organization* (ISO)/*open systems interconnection* (OSI) reference model] but there was no decision for the transmission type—whether it would be analog or digital [1, 2].

Today the GSM standards number over 8,000 pages, and their development and maintenance is the responsibility of the European Telecommunications Standards Institute (ETSI). The transition from CEPT to ETSI occurred in 1989, and from 1991 to 1999 GSM standardization was the responsibility of the Special Mobile Group (SMG) within ETSI. Since 1999, GSM specifications have been combined with the 3G specifications under the auspices of the 3G Partnership Project (3GPP) [3].

3GPP is a collaboration agreement that was established in December 1998. The original scope of 3GPP was to produce globally applicable technical specifications and technical reports (TRs) for a 3G mobile system based on evolved GSM core networks and the radio access technologies that they support [i.e., UMTS terrestrial radio access (UTRA) both frequency division duplex (FDD) and time division duplex (TDD) modes]. The scope was subsequently amended to include the maintenance and development of the GSM technical specifications and TRs including

---

1. This was the original meaning of GSM.

evolved radio access technologies [e.g., general packet radio service (GPRS) and enhanced data rates for GSM evolution (EDGE)].

Five technical specification groups (TSGs), namely the services and system aspects (SA), core network (CN), terminals (T), radio access network (RAN), and GSM EDGE RAN (GERAN) groups, carry out the technical work. The main groups within 3GPP together with their subgroups; their specific tasks are depicted in Figure 2.1.

The combined 3G and GSM specifications are divided into 15 series as indicated in Table 2.1. They have a 3GPP specification number consisting of four or five digits (e.g., 01.01 or 21.001). The first two digits define the series as listed in Table 2.1, and they are followed by two further digits for the 01 to 13 series (GSM only) or three further digits for the 21 to 55 series (3G and GSM) that define the version of the respective specification.

A specification in the 21 to 35 series may apply either to 3G only or to GSM and 3G. This can be determined by the third digit, where a "0" indicates that it applies to both systems. For example, 21.$\underline{0}$01 applies to 3G and GSM systems, whereas 21.$\underline{1}$01 and 21.$\underline{2}$01 apply only to 3G.

The 3GPP specifications are evolving continuously, and new features are supported to meet emerging market requirements. To provide developers with a stable platform for implementation while at the same time allowing for the addition of new

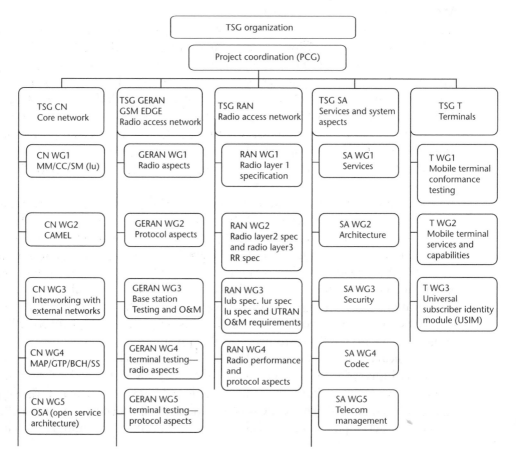

Figure 2.1   The organization of the technical specification groups of 3GPP.

Table 2.1 Number and Subject of the 3GPP Specifications

Subject of Specification Series	3G/GSM R99 and Later	GSM Only (Rel-4 and Later)	GSM Only (Before Rel-4)
Requirements	21 series	41 series	01 series
Service aspects	22 series	42 series	02 series
Technical realization	23 series	43 series	03 series
Signaling protocols (user equipment to network)	24 series	44 series	04 series
Radio aspects	25 series	45 series	05 series
CODECs	26 series	46 series	06 series
Data	27 series	47 series	07 series
Signaling protocols (RSS-CN)	28 series	48 series	08 series
Signaling protocols (intrafixed-network)	29 series	49 series	09 series
Program management	30 series	50 series	10 series
User identity module (SIM/USIM)	31 series	51 series	11 series
O&M	32 series	52 series	12 series
Access requirements and test specifications	—	13 series	13 series
Security aspects	33 series	—	—
Test specifications	34 series	—	11 series
Security algorithms	35 series	55 series	—

features, the 3GPP uses a system of parallel "releases," all of which are listed in Table 2.2. The list of specifications needed for each 3G and GSM release is listed in the TRs, which are also mentioned in Table 2.2.

The full title, specification number, and latest version number for every specification can be found in the status list [4], and more information about the specification releases can be found on the releases and phases Web page [5].

The remainder of this chapter deals only with GSM specifications and in particular with Release 4, which is considered to be the most widely implemented today (early 2003).

2.1.1 Functional Architecture of GSM

The GSM system is one of the most complex systems that humans have devised, and this justifies the approximately 10 years of time that was needed to successfully put

Table 2.2 3GPP Specifications as of March 2003

GSM/EDGE Release	3G Release	Abbreviated Name
Phase 2+ Release 6 (TR 41.104)	Release 6 (TR 21.104)	Rel-6
Phase 2+ Release 5 (TR 41.103)	Release 5 (TR 21.103)	Rel-5
Phase 2+ Release 4 (TR 41.102)	Release 4 (TR 21.102)	Rel-4
—	Release 2000	R00
Phase 2+ Release 2000	—	
—	Release 1999 (TR 21.101)	R99
Phase 2+ Release 1999 (TR 01.01)	—	
Phase 2+ Release 1998	—	R98
Phase 2+ Release 1997	—	R97
Phase 2+ Release 1996	—	R96
Phase 2	—	Ph2
Phase 1	—	Ph1

it into operation. It consists of several functional entities or subsystems that extensively exploit advanced computer and communications technologies. The GSM subsystems are grouped together into three main entities, namely the radio subsystem, the network and switching subsystem, and the operation and maintenance subsystem, which communicate, over specific interfaces. Figure 2.2 depicts the structure of each subsystem as well as the interfaces through which communication is accomplished.

2.1.2 Radio Subsystem

The main entities within the radio subsystem are the air interface and base station subsystem (BSS). The BSS consists of the base transceiver stations (BTSs) and of the base station controller (BSC). The next sections detail the structure and functionalities of each one of the main parts of the radio subsystem.

2.1.3 Air Interface

The air interface (or radio interface) is positioned physically between the MS and the BTS. The reference point corresponding to this interface is called U_m. This section briefly describes concepts related to the radio aspects corresponding to this interface.

The core of the specifications for the air interface are provided in series 45 (radio aspects) of the GSM specifications. This series consists of 11 different documents as described in Table 2.3. They specify how the mobile stations (MSs) communicate with the BTS.

The MS is the means through which the user accesses the services offered by the available cellular networks. The MS as a device consists of two parts. The first part corresponds to the hardware and software components that enable binding to the air interface, while the second part corresponds to the hardware and software components that handle and maintain the user's personal data and preferences.

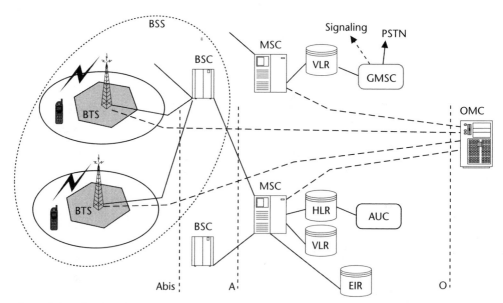

Figure 2.2   The GSM functional architecture.

Table 2.3 Contents of the 45 Series Specifications Defining the GSM Air Interface

45.001	"Physical Layer on the Radio Path (General Description)"
45.002	"Multiplexing and Multiple Access on the Radio Path"
45.003	"Channel Coding"
45.004	"Modulation"
45.005	"Radio Transmission and Reception"
45.008	"Radio Subsystem Link Control"
45.009	"Link Adaptation"
45.010	"Radio Subsystem Synchronization"
45.022	"Radio Link Management in Hierarchical Networks"
45.050	"Background for RF Requirements"
45.056	"CTS-FP Radio Subsystem"

The radio subsystem provides a certain number of logical channels that can be separated into two categories, the *traffic channels* (TCHs) and the *control channels*. The structure of these channels is defined in [6].

Traffic channels carry two types of user information streams: encoded speech and data. The following types of traffic channels are defined: B_m or full-rate (TCH/F), L_m or half-rate (TCH/H), and the cell broadcast channel (CBCH). The gross rate for TCH/F is 22.8 Kbps while the gross rate for the TCH/H is 11.4 Kbps. These types of channels can appear in various forms depending on the applications' requirements or the infrastructure's constraints. For example the *full-rate data TCH* can appear in forms that can accommodate traffic from 4.8 Kbps up to 43.2 Kbps and is distinguished when mentioned using different lettering (e.g., TCH/F4.8 and TCH/F9.6). The set of traffic channels that are defined for speech and for data according to Release 4 of the GSM specifications are given in Table 2.4. Packet data traffic channels are not listed since they are out of the scope of this chapter.

Control channels are intended to carry signaling or synchronization data. Control channels are subdivided into broadcast control channel (BCCH), common control channel (CCCH), stand-alone dedicated control channel (SDCCH), and associated control channel (ACCH). An ACCH is always allocated in conjunction with either a TCH or an SDCCH. Two types of ACCHs for circuit-switched

Table 2.4 Traffic Channels as Defined in Release 4 of GSM Technical Specification

Traffic Channel Name	Codeword Used
Full-rate traffic channel for speech	TCH/FS
Half-rate traffic channel for speech	TCH/HS
Enhanced full-rate traffic channel for speech	TCH/EFS
Adaptive full rate traffic channel for speech	TCH/AFS
Adaptive half-rate traffic channel for speech	TCH/AHS
Full-rate traffic channel for 9.6-Kbps user data	TCH/F9.6
Full-rate traffic channel for 4.8-Kbps user data	TCH/F4.8
Half-rate traffic channel for 4.8-Kbps user data	TCH/H4.8
Half-rate traffic channel for 2.4-Kbps user data	TCH/H2.4
Full-rate traffic channel for 2,4-Kbps user data	TCH/F2.4
Full-rate traffic channel for 14.4-Kbps user data (TCH/F14.4);	TCH/F14.4
Enhanced circuit-switched full-rate traffic channel for 28.8-Kbps user data	E-TCH/F28.8
Enhanced circuit-switched full-rate traffic channel for 32.0-Kbps user data	E-TCH/F32.0
Enhanced circuit-switched full-rate traffic channel for 43.2-Kbps user data	E-TCH/F43.2
Cell broadcast channel	CBCH

connections are defined: continuous stream (slow ACCH) and burst stealing mode (fast ACCH). As in the traffic channels, control channels appear also in various forms depending on the applications' requirements or the infrastructure's constraints. In Release 4 of the GSM technical specifications, four categories of control channels were defined:

- BCCHs;
- CCCHs;
- Dedicated control channels;
- Cordless telephony system (CTS) control channels.

The set of control channels that are defined for speech and for data according to Release 4 of GSM specifications are given in Table 2.5. Packet control channels and CTS control channels are not listed since they are out of the scope of this chapter.

The BCCH is used to transmit information from the base stations to the mobile stations regarding the status of the network and the frequencies being used as well as regarding the existence of certain features like frequency hopping.

The FCCH is used to inform the mobile station about the frequency being used from the BTS for the MS to correct its transmitting frequency if such an action is required.

The SCH transmits synchronization bursts to the MS to help it to be aligned in time with the transmission from the BTS.

The PCH is used to find and address the MS when a call is initiated. It exists only in the direction from the BTS to the MS.

The RACH is used from the MS to request channel capacity from the BTS. For that purpose a slotted ALOHA access protocol is being used.

Table 2.5 Control Channels as Defined in Release 4 of the GSM Technical Specification

Category	Control Channel Name	Codeword Used
BCCHs	BCCH	BCCH
	Frequency correction channel	FCCH
	Synchronization channel	SCH
Common control channels	Paging channel	PCH
	Random access channel	RACH
	Access grant channel	AGCH
	Notification channel	NCH
Dedicated control channels	Slow, TCH/F, or E-TCH/F-associated, control channel	SACCH/TF
	Fast, TCH/F-associated, control channel	FACCH/F
	Slow, TCH/H-associated, control channel	SACCH/TH
	Fast, TCH/H-associated, control channel	FACCH/H
	SDCCH	SDCCH/8
	Slow, SDCCH/8 associated, control channel	SACCH/C8
	SDCCH, combined with CCCH	SDCCH/4
	Slow, SDCCH/4 associated, control channel	SACCH/C4
	Slow, TCH/F or E-TCH/F associated, control channel for multislot configurations	SACCH/M
	Fast, E-TCH/F associated, control channel	E-FACCH/F
	Inband, E-TCH/F associated, control channel	E-IACCH/F

The AGCH is used to respond to an incoming request over the RACH with the allocation of an SDCCH or a TCH.

The SDCCH is used for the transmission of control information like registration and location update. It is a two-way channel between the MS and the BTS and is always used when a traffic channel has not been assigned.

The SACCH is always allocated in parallel with an SDCCH or a TCH and is used for the transfer of parameters that help to assess the communication quality.

Finally, the FACCH is used when a TCH is existent and is usually used when a handover process is going to happen to ensure a transition that is as smooth as possible.

Logical channels are mapped into physical channels as also defined in [6]. The available radio spectrum is partitioned in both frequency and time. Frequency is partitioned into 200-KHz bandwidth channels, and each one of them is further divided into eight physical TDM channels, which consist of eight timeslots of a 0.577-ms duration. Eight timeslots form a TDMA frame with a duration of 4.62 ms, which corresponds to the basic multiplexing structure in GSM. Thus a physical channel is characterized by both its carrier frequency and the timeslot available to it, which appears every 4.62 ms. Over the time -slots data is transmitted in the form of *bursts* that have a length of 148 bits. If a message is longer than 148 bits it has to be split up into several bursts before transmission.

For each logical channel a specific coding and interleaving scheme has been defined for achieving maximum efficiency. Thus different channel coding methods are applied on the various channels for the different services. However, special attention was paid to the coding and interleaving procedures to organize them in such a way to allow a homogeneous decoder structure.

The following sequence of actions takes place for each channel before transmission:

- The bits of information are coded with a systematic block code resulting in words of information plus parity bits;
- This information plus parity bits are encoded with a convolutional code resulting in the coded bits;
- The coded bits are reordered and interleaved, and with the addition of a stealing flag we acquire the interleaved bits.

These actions are always performed over a block of bits, which is different from channel to channel and has a minimum value of 36 coded bits and a maximum value of 456 coded bits, which is the most common size used for a block.

The modulation format that is used by the standard GSM systems is a digital Gaussian minimum-shift keying (GMSK). EDGE systems are using the phase shift keying (8PSK) modulation to achieve higher data rates [7].

Frequency bands and channel arrangement, transmitter and receiver characteristics, as well as transmitter and receiver performance are described in detail in [8]. Other details regarding the radio link such us power control, adaptation and synchronization are provided in [9–11]. Their study is out of the scope of this book, and the same applies for the speech coding techniques. However, readers who would like to learn more about these issues can visit the 3GPP Web site and the associated ftp server where they can download all the latest documentation.

The remainder of this section focuses on the signaling related to the radio interface that is required for the realization of the MS-BSS communication. The MS-BSS interface from the signaling point of view is specified in the 44-series of the 3GPP technical specifications.

For the accomplishment of signaling functions on the MS-BSS interface three layers are required as depicted in Figure 2.3 according to the OSI model as it is given in ITU-T recommendation X.200 [12].

The *physical layer* is the lowest layer, and its role is to support the transfer of bit streams over the radio medium. It includes error detection and correction functions as well as encryption to support an error-free and secure transmission. The specific services that this layer offers to the MS are: (1) establishment of dedicated physical channels, (2) establishment of packet data physical channels, and (3) cell/PLMN selection in idle mode or in packet mode. The functions and protocols of the physical layer are defined in a technical specification [13].

The *data link layer* defines the frame structure and the procedures for peer-to-peer communication. The functions and protocols of the data link layer are defined in technical specifications [14, 15].

Layer 3 is responsible for a number of very important functions that make mobile telephony a reality. Particularly this layer is responsible mainly for supporting RRM, for mobility management (MM), and for the connection management (CM) functions. The functions and protocols of layer 3 of GSM are defined in technical specifications [16, 17].

2.1.4 BSS

As we have already mentioned, the BSS is comprised of the BTS and the BSC. The BSS incorporates all the radio-related capabilities and characteristics that were described in the previous section. The mobile stations communicate with the rest of the GSM network through the BTS part of the BSS via the air interface. The two main elements of the BSS communicate with each other over the standardized A_{bis} interface.

The main parts of the BTS are the transmitting and receiving equipment and the facilities that handle the signaling over the air interface. Other important elements that are contained in the BTS are the TRAU, speech coding and decoding components, as well as rate adaptation mechanisms for data transmission. Depending on the type of the antenna used a BTS may support one or several cells.

The BSC is responsible for the management of the radio resources for one or more BTSs. It handles the reservation and release of radio channels and manages

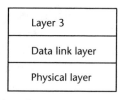

Figure 2.3 Layering on the MS-BSS interface.

handovers. The BSS is connected to the GSM fixed network over the so-called A-interface, which connects the BSC with the mobile switching center (MSC).

2.1.5   Network and Switching Subsystem

All the network-related functions and the packet switching are carried out by the network and switching subsystem (NSS), which forms the gateway between the radio network and the associated fixed network [e.g., public switched telephone network (PSTN) and ISDN]. The main components of the NSS are the MSC, the home location register (HLR) and the visitor location register (VLR).

The MSC is responsible for handling the connection to and from the fixed networks. It operates like a normal switching node of the associated PSTN or ISDN network, and it is responsible for all the signaling required for setting up, maintaining, and terminating connections, in accordance with common channel signaling system no. 7 (SS7). Each MSC is usually allocated several BSCs and provides all the functionality needed to handle a mobile subscriber, such as registration, authentication, location updating, handover management, and call routing to a roaming subscriber. Other tasks of the MSC include the well-known ISDN supplementary services like call forwarding, call baring, conference calling, and call charging to the user called.

The HLR contains all the administrative information such as telephone number, MS identification number, equipment type, subscription basis and supplementary services, access priorities, and authentication code for each subscriber registered in the corresponding GSM network. All this data is referred to as quasi-permanent static data. The current location of the MS and the mobile station roaming number (MSRN), which is also necessary for setting up a connection, is called temporary subscriber data and is also stored in the HLR. This information is immediately updated whenever the mobile subscriber changes location area (LA). Each mobile subscriber's data is registered in only one HLR where billing and other administrative tasks are performed. There is usually one HLR per GSM network, although it is often implemented as a distributed database.

The role of the VLR is to manage the subscribers who are currently roaming, and it thus contains selected administrative information from the HLR, necessary for call control and provision of the subscribed services, for each mobile currently located in the geographical area controlled by the VLR. A VLR is always under the control of an MSC, and that is why many times they are manufactured as a single functional entity. If the mobile user traverses several of the MSC's location areas, the VLR data that is related to this subscriber should be updated and the same happens in the event of an MSC change.

2.1.6   The Operation Subsystem

All the operation and maintenance functions are performed by the operation subsystem, which consists of three main elements: (1) the operation and maintenance center (OMC), (2) the authentication center, and (3) the equipment identity register.

The OMC monitors and controls the rest of the network elements over the standardized O-interface and guarantees the best possible service quality for the user.

The management functions of the OMC include administration of subscribers and equipment, generation and processing of statistical data, billing, and alert handling whenever unexpected events occur.

In the operation subsystem (OSS) there are also two databases that are used for authentication and security purposes: the equipment identity register (EIR) and the authentication center (AuC). The EIR is a database that contains a list of all valid mobile equipment on the network, where each MS is identified by its international mobile equipment identity (IMEI). An IMEI is marked as invalid if it has been reported stolen or is not type-approved. The AuC is a protected database that stores a copy of the secret key stored in each subscriber's SIM card, which is used for authentication and encryption over the radio channel.

The functions of the OSS are subdivided into three areas, listed as follows:

1. Subscription management;
2. Mobile equipment management;
3. Network operation and maintenance.

Subscription management is responsible for authenticating the mobile subscribers from the data that are stored in the HLR and for service billing. The billing location can be the MSC in which the mobile user is currently active or a gateway MSC (GMSC). The GMSC is the interface between the cellular network and the PSTN. The PSTN sends all calls to the GMSC, which finds the right MSC, which connects to the mobile.

Mobile equipment management is responsible for storing and handling information regarding the owner of mobile equipment together with its associated identity.

Network operation and maintenance concerns the control of the network elements based on the TMN management concepts as they have been developed by the ITU-T. The TMN functions are divided into four categories: (1) business management, (2) service management, (3) network management, and (4) network element management.

2.1.7 Interfaces of the GSM System

Except for the radio interface, which was explained in detail in the previous sections, the following important interfaces exist in the GSM systems:

- The BTS-BSC interface at reference point A_{bis}, which is at the physical layer a PCM link at 2,048 Kbps with a frame structure of 32×64-Kbps time slots. The A_{bis} interface has two types of communication channels: traffic channels at 8, 16, or 64 Kbps carrying speech or data of one radio traffic channel (full-rate or half-rate), signaling channels at 16, 32, or 64 Kbps, carrying signaling information 3GPP TS 48.051 [18] and 3GPP TS 48.052 [19].
- The BSS-MSC interface at reference point A, which at the physical layer consists of one or more PCM links at 2,048 Kbps based on CCITT Rec. G703 3GPP TS 48.001 [20] and 3GPP TS 48.002 [21].
- The BSC/MSC-OMC interface at reference point O, which is based on recommendation X.25 of ITU-T [22].

2.2   Network Performance Evaluation of 2G Systems

This section presents an in-depth performance evaluation study performed in the network of a cellular operator. This study is important since it highlights the congestion problems on the air interface of 2G wireless systems. For that purpose we briefly present the reporting mechanism in GSM, since the performance evaluation is based on this kind of measurement. Subsequently, network behavior in normal situations is presented, and finally, traffic overload scenarios are discussed. At this point it has to be mentioned that the information and diagrams presented in the following sections are not selected randomly, but from BSC areas that suffer the most from traffic congestion. However, they clearly describe the behavior of a GSM system, in terms of performance degradation independently from the cellular operator.

2.2.1   Cellular Network Reporting

The performance of cellular networks is the most important issue for the operators, from a technical point of view. Their main goal is to keep the subscribers satisfied with the QoS they provide. To achieve the best performance, they have to monitor and optimize their network continuously. A NMS with an on-line database is responsible for the collection of every report generated in the network, in a raw data form. For greater effectiveness, operators install systems that do more than collect and store (e.g., they organize data in event counters and generate Web reports). In Section 2.2.1.1 such a data warehouse system, which has been studied for its historical data, is presented.

2.2.1.1   DW

The DW is deployed to assist the optimization the performance of telecommunication networks. It collects, stores, manages, and presents data on a long-term basis. The DW is a combined data collection and data handling system comprised of several individual servers. DW servers are responsible for data collection and storage, preprocessing tasks, report generation, data analysis, and report publishing via the operator's intranet. The cellular operator can utilize the DW to automatically generate Web page–based reports that are created from scratch and reports requested according to given parameters [23].

The DW stores data for longer periods of time as opposed to the NMS on-line database. For that reason part of DW is located on a separate database server that allows storing performance measurements, alarms, and radio planning network parameters for a long period of time. It can store this data for several years. Raw performance measurement (PM) data, alarm/fault management (FM) data, and radio network parameter (RNW) data is extracted from the operational NMS database and stored periodically in the DW. Data is processed according to the different types. Figure 2.4 shows the procedure of collecting and processing data in DW.

2.2.1.2   Optimizing NMS with DW

The load on the network management system depends on the size of the network and the frequency and effectiveness of measurements. By keeping daily on-line

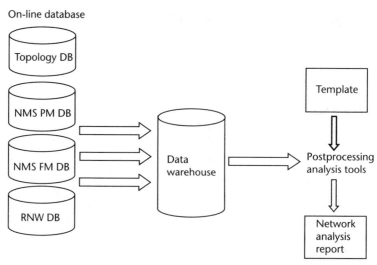

Figure 2.4   Collecting and processing data in DW.

processing separated from the off-line analysis of long-term data, DW ensures an improved overall performance of the NMS. Summarizing raw PM data in DW, leads to the reduction of the load of the NMS, as well as to storing measurement data for substantially longer periods of time. Stored data can be used in DW to help forecasting of future events and trends concerning the traffic in the network. In the DW, FM, PM, and RNW data is reorganized and combined together in such a way to be easier to generate trend data reports that are as useful as possible.

Long-term data in DW can be extracted and processed with NMS post-processing tools designed for PC environments. These tools enable the presentation of long-term data in various reports, which show trends in the functioning of the network, the behavior of end users, as well as relations between the performance measurements data, alarm data, and radio network parameter data.

2.2.2   Statistical Evaluation in GSM

This section of the document deals with the presentation of the measurements performed in a cellular network. The most reliable and representative performance indicators have been chosen throughout the analysis of operational data. To identify a congestion situation, this section refers to patterns of a normal day, while the following section refers to congestion situations. The performance indicators presented are traffic, call set up success rate (CSSR), handover success rate (HOSR), SDCCH blocking rate (BR), and TCH BR [23–25].

Traffic is chosen to estimate the demand for services and channels in actual numbers. CSSR and HOSR are chosen to appreciate the impact of congestion in the two most important procedures during a call attempt, regarding the QoS offered to the subscribers. Finally, SDCCH and TCH BRs are chosen to analyze how logical channels are affected when congestion appears, since the representative channels are the most affected in a congestion situation.

In this section we present a number of diagrams that show the network's performance in a normal situation in order to use them as patterns for the expected network's behavior. Traffic, CSSR, HOSR, SDCCH BR, and TCH BR are the indicators used. On a normal day their behavior is examined by averaging the measurements on BSC level every hour of the day. It must be noted that days that include high congestion events are excluded from the averaging process, as they do not comply with the normal day pattern. In addition, these data may vary from operator to operator, or even from area to area. In addition, the data are selected from a BSC that represents the average performance of a cellular GSM network.

The particular pattern of traffic in Figure 2.5 presents two peaks: a lower one around 1 P.M. and a higher one around 7 P.M. Usually, busy hour is considered to be at noon, but our data analysis showed higher values in the afternoon peak. In addition, traffic reaches minimum values of below 100 Erlang during the night (for a singe BSC area), until the early morning hours. This pattern repeats itself every day with a slight variation on weekends (and especially on Sundays) where the noon peak is much lower than the usual.

In Figure 2.6 the CSSR pattern shows an inverted link with the one of the traffic. CSSR reaches minimum points at the exact time that traffic reaches its peaks, showing that the increase of traffic degrades this particular success rate. The difference between the two charts is during the night, where CSSR also reaches minimum values.

The HOSR pattern in Figure 2.7 is an exact duplicate of the traffic pattern, but also inverted. It shows the lowest values when traffic is at its highest ones, while the minimum HOSR coincides with the traffic peak, thus showing the existence of a chain effect phenomenon.

The SDCCH BR pattern in Figure 2.8 is also similar to the traffic pattern presenting the same peaks at 1 P.M. and 7 P.M. and minimums at night, showing that increased traffic increases blocking too.

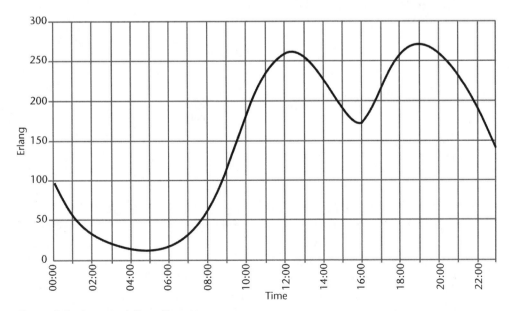

Figure 2.5   Average daily traffic pattern.

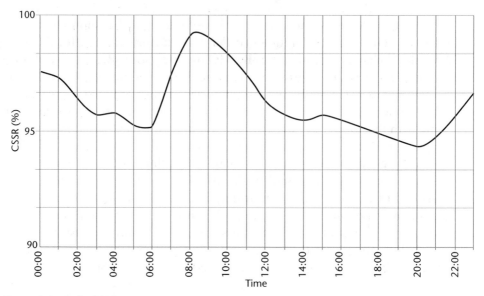

Figure 2.6 Daily CSSR pattern.

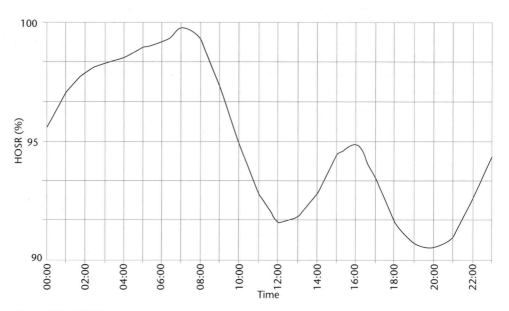

Figure 2.7 HOSR pattern.

The TCH BR pattern in Figure 2.9 is exactly the same pattern as the traffic one. One can notice the two peaks of the TCH BR during the same period with the accommodated traffic in Erlang.

Looking carefully at Figures 2.5–2.9, it is obvious that all indicators depend on traffic, which affects their behavior and those of all the procedures that these counters represent. This so-called chain effect phenomenon is more clearly perceived in Section 2.2.3, where congestion charts are presented.

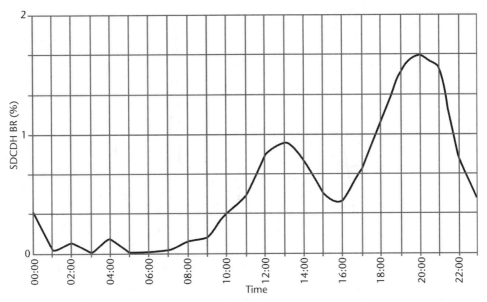

Figure 2.8   Daily SDCCH BR pattern.

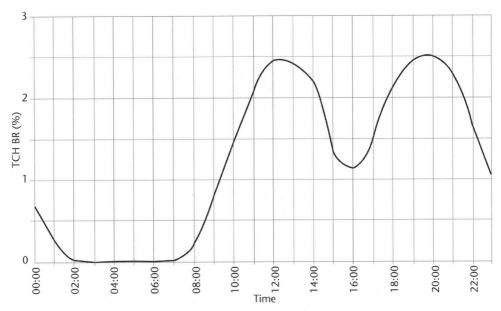

Figure 2.9   TCH BR pattern.

2.2.3   Evaluation of Operational Data in Traffic Overload Situations

Throughout the whole procedure of data processing and analysis, the chain effect phenomenon was encountered in every possible congestion situation. As the study focused on the most important performance indicators, a strong relation between them was observed in every case we examined. The CSSR, HOSR, TCH BR, and SDCCH BR were studied with the help of the traffic counter. These indicators and counters are properly defined and graphically presented in Figures 2.10–2.14.

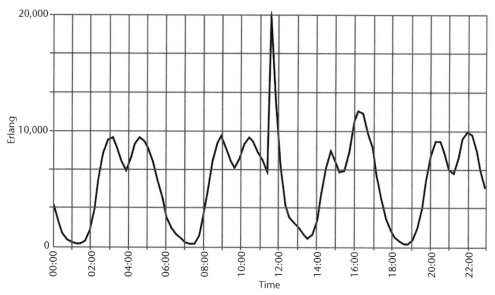

Figure 2.10 Traffic in a high congestive situation.

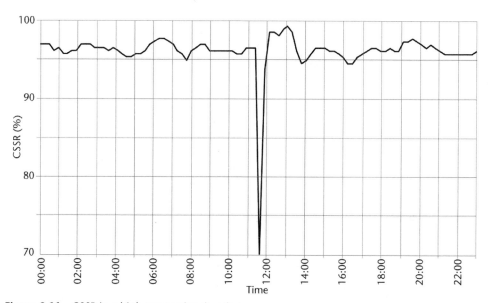

Figure 2.11 CSSR in a high congestive situation.

As the data was thoroughly analyzed, both at a network and BSC level and for large- as well as low-scale events, we observe that these counters were following one another; in other words, one was triggering the other. A rapid increase of traffic caused a serious increase in the TCH and SDCCH BRs, thus triggering dramatic degradation of CSSR and HOSR. As the study was extended to other counters and indicators, it was observed that they followed the same pattern.

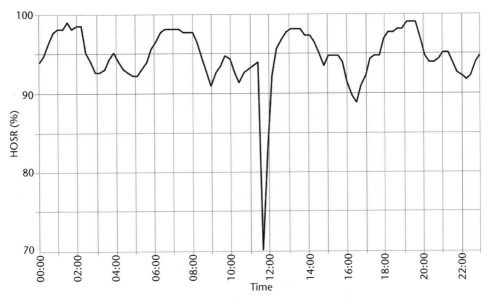

Figure 2.12   HOSR in a high congestive situation.

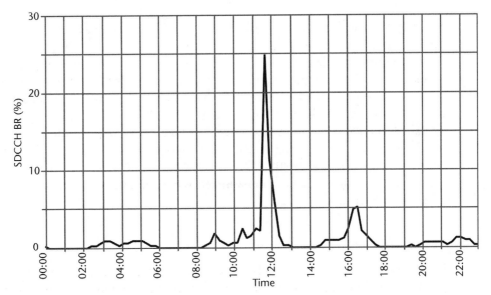

Figure 2.13   SDCCH BR in a high congestive situation.

2.2.3.1   Large-Scale Event

Under large-scale events we focus on high-congestion situations with a relatively long duration and generally a large geographical size (e.g., New Year's Eve). Figures 2.10–2.14 depict such a congestion situation, where a large part of the network was affected.

A predictable and expected event, such as the New Year's Eve situation, is depicted in Figure 2.10. Traffic increased about 220% over a busy hour (BH) and

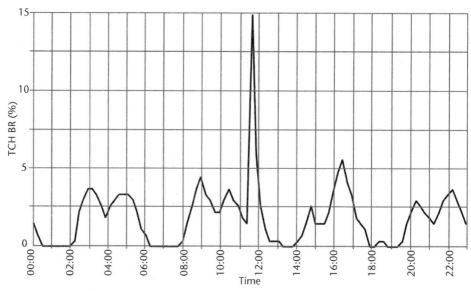

Figure 2.14 TCH BR in a high congestive situation.

about 400% over the same hour of a normal day. The event was perceived by a large part of the entire network.

On the network level, the TCH BR responds to the congestion situation reaching the value of 15%, while on a normal day BH is around 2%. SDCCH BR also perceives the heavy load of traffic as it climbs to 25%, while in normal situations it is about 2% (up to 5% in extremely loaded cells). As described above, the CSSR is degraded to 70% from the usual 95 to 98% that is observed. At the same time, the handover procedure is also affected, as the HOSR falls to 72%, while in normal cases it reaches values over 90%. On a BSC level, as all BSCs of the network contribute to the congestion situation on New Year's Eve, the chain effect has, more or less, a similar behavior.

2.2.3.2 Low-Scale Event

Under low-scale events we classify all congestion situations that are not severe, compared to the ones presented in the previous section. The affected cells are less, and the duration is most of the time shorter.

In Figure 2.15 a congestion event is perceived by only a part of the network (geographically limited), and it is neither predicted nor expected (local earthquake; only few BSCs participate). Traffic increases by around 10% on a network level during such events.

The congestion situation that is presented in Figures 2.16–2.19 has occurred due to a small earthquake in the corresponding area and was strongly linked to an increased number of incoming and outgoing call requests. On a network level, TCH BR reaches 7.5%, while the day before and the day after, at the same time, it was 2.4%, and the usual rate is 3.25% during a typical BH. On a BSC level, BSCs that do not have cells near the epicenter of the earthquake do not perceive the phenomenon,

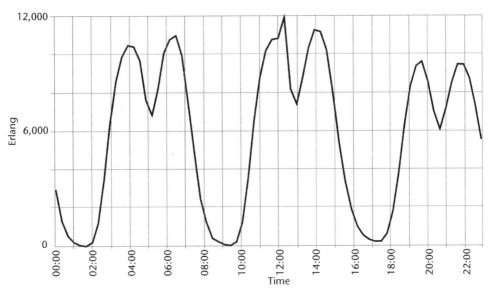

Figure 2.15 Traffic in a low congestive situation.

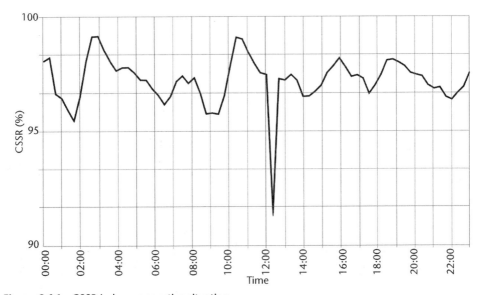

Figure 2.16 CSSR in low congestive situation.

while BSCs near the epicenter reach rates of 15 to 24%. SDCCH BR climbs to 16% on a network level, many times higher than the usual rate, but in the limits of a normal busy hour. On the BSC level, only BSCs near the epicenter are affected, climbing to rates of 3% up to 5%. As described above, the CSSR is triggered, and a degradation from 97% to 92% occurs on the day before and the day after. On the BSC level, only BSCs near the epicenter are affected, climbing to rates of 70 to 85%. At the same time, the handover procedure is also affected, as the HOSR falls to 83%, while in normal cases it reaches values between 90% and 98% (as it was the day before

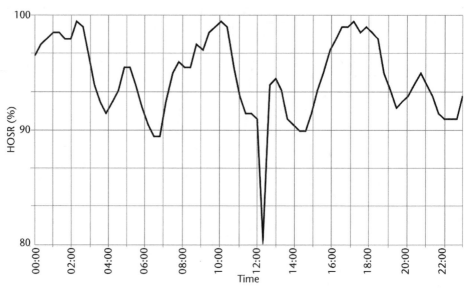

Figure 2.17 HOSR in low congestive situation.

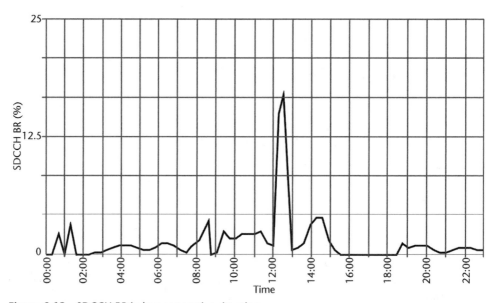

Figure 2.18 SDCCH BR in low congestive situation.

and the day after, at the same time). Moreover, on the BSC level, the results are even more persuasive for the chain effect phenomenon. Randomly chosen BSCs with excellent behavior in one of the KPIs present excellent behavior in the other KPIs too. Randomly chosen BSCs with average behavior in one of the KPIs present the same results in the other KPIs too. The same thing happens to BSCs with bad behavior respectively. There are of course exceptions in that particular analysis, which are due mostly to other external reasons.

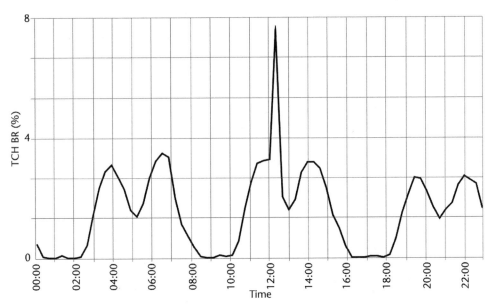

Figure 2.19   TCH BR in low congestive situation.

2.2.3.3   Radio Network Performance in a Congestion Situation

Apart from the basic performance indicators, the study was extended to different counters to achieve a better overview of the network's behavior. Measurements from the A_{bis} interface show that a heavy traffic load, though it can be handled by the air interface, presents a bottleneck in the A_{bis} point, concerning mostly SDCCH BR, paging procedure, handover performance, and less TCH BRs. A look at handover-related counters confirm that fact, as directed retry handovers climb to 13% in a large-scale situation and 7% in a low-scale case, while in normal cases they reach only 2%. That shows how the lack of resources affects the handover procedure to a greater extent than the HOSR. Congestion duration in both TCH and SDCCH was also investigated. Measurements show that SDCCH congestion duration was lower than 30 sec on normal days, while it reached 1–4 min to the BSCs affected by the earthquake that day and climbed to 4–7 min in the large-scale event. TCH congestion duration followed the same trend with lower rates, but in the earthquake event this did not decline from the usual rates. Therefore, only SDCCH encountered problems that day and not the traffic channels. Finally, occupation time of the SDCCH was analyzed and compared to SDCCH BR, for all the above situations, on the BSC level. The results show that the increase of the BR is followed by a respective increase of the SDCCH occupation time, no matter which channel is used, no matter what the situation is. Figure 2.20 shows exactly this for a randomly chosen BSC.

Therefore, the delay observed in successfully sending an SMS during a congestion situation occurs due to the heavy congestion, and this delay makes the problem even worse, since it occupies the channel for a longer period of time. Obviously this delay is observed by all users trying to set up a call or send an SMS message, something that dramatically increases the congestion of the SDCCH.

Finally, after careful processing and analyzing the operational data given and evaluating the results produced, the most important conclusion is that the

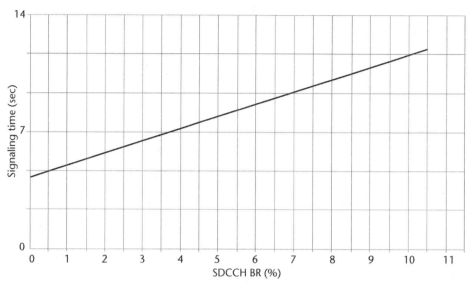

Figure 2.20 Occupation time versus SDCCH BR.

congestion problem affects a greater part of the network, even if it is low-scaled or locally placed. It cannot be limited to one network element or to one procedure, but, as shown above, it causes traffic overload in the air- and the A_{bis} interface, in most of the logical channels used, in the call set up and the handover procedures by a chain effect phenomenon. Though SDCCH encounters most problems—not only in the air interface but also in the A_{bis} and therefore should be carefully treated—the congestion problem should be dynamically confronted in a way that allows solutions to apply to the network level.

2.2.3.4 System Performance at the BSC Level

Figure 2.21 presents one of the major observations during our studies. The analysis has been focused on cells of a BSC, which handled an increased traffic volume of a specific traffic load scenario. The requested traffic in the area during the event caused serious performance degradation of the network. The black line indicates the SDCCH BR, while the gray bars indicate the SDDCH requests. One can clearly see that the system responds quite well in all situations, no matter the amount of SDCCH requests and that the BR is almost constant and close to 0%. Once the system reaches a specific threshold, the SDCCH BR increases dramatically and is close to 35%.

This is another proof that shortcomings of the SDCCH due to A_{bis} failures contribute to the extended reservation time of the SDCCH channels, affect the overall resource availability, and increase further the BR.

In Figure 2.22, the network performance during the same event is depicted. During this time the drop call rate (DCR) is shown with a red line, while the blue bars indicate the SDCCH requests. Again, the system performance is very good, as long as a threshold is reached. After that the DCR increases dramatically and gets close to 90%!

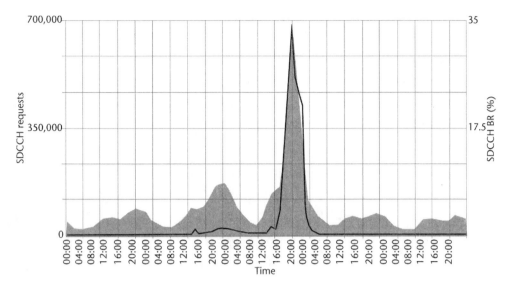

Figure 2.21   SDCCH blocking chart.

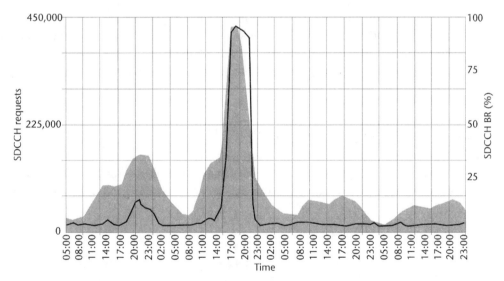

Figure 2.22   SDCCH DCR.

Figure 2.23 depicts the TCH blocking according to the call requests. Once again the system can accommodate the offered traffic, as long as the requests do not reach the instability threshold. After the system reaches that, the BR increases dramatically and is more than 30%.

In Figure 2.24, the increase of DCR is depicted, while call requests reach the threshold. This can be observed in all congestion situations, where the blocking probability and all other (chain) effects are increasing rapidly after a specific value.

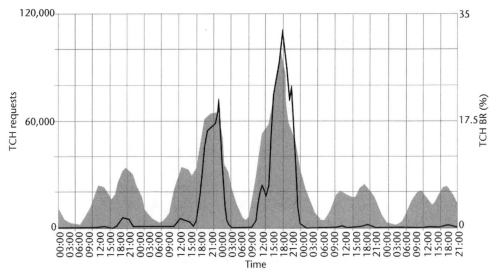

Figure 2.23   TCH blocking—TCH call requests.

Finally, the HOSR, as depicted in Figure 2.25, shows that the chain effect influences the handover performance, since handover requests cannot be handled due to the traffic overload in the target cell. This observation is very important, since it dictates that there should be a mechanism to avoid the handover in an already congested cell.

2.2.3.5   System Performance at the BTS Level

This section extends the study to the BTS level. In that way, the evaluation of operational data is performed on a deeper level, reaching the fundamental network element, the cell. The situation analyzed is an event that is considered as a predictable,

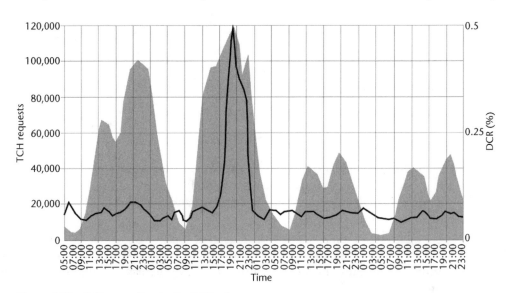

Figure 2.24   DCR according to the TCH seizures.

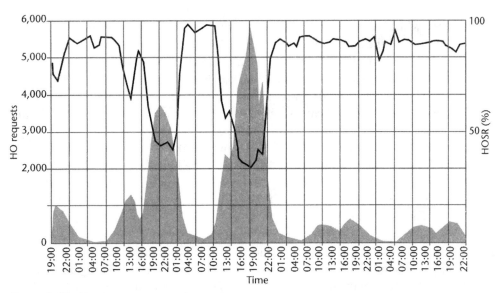

Figure 2.25   Handover success performance.

high congestive one, on a cell basis, though at a network level is a low congestive one or not even perceived by the network.

The event was a very popular football match, held in a stadium of high capacity, located inside an urban area. The area is mainly served by three cells (STD_A, B, and C). The event gathered about 70,000 people in the stadium. The percentage of this figure that is subscribed to the specific cellular operator shows exactly the extra load that the three cells had to serve, in addition to the normal traffic. The study focuses on the most representative performance indicators: CSSR, HOSR, SDCCH BR, TCH BR, and traffic that are used in most of the situations presented in this chapter. Data is analyzed not only for the date of the event, but for the day before and the day after, to compare the results. Table 2.6 provides all the data mentioned above, with the daily column referring to the mean values for the day.

It should be noted that the situation started at 8 P.M. and lasted until 12:30 A.M. That affected the definition of busy hour of each day. On the day of the event, the busy hour is defined at 8 P.M., while the day after, it is 12 A.M. (the first hour of the

Table 2.6   BTS Statistics

Cell ID	Date	CSSR (%) Daily	BH	HOSR (%) Daily	BH	SDCCH BL(%) Daily	BH	TCH BL (%) Daily	BH	Traffic (Erl) Daily	BH
STD_A	Day before	92.9	100	95.8	100	0	0	0	0	0.222	1.626
	Event day	32.7	12.7	46.6	36.4	26.6	27.1	48.9	71.9	4.509	27.34
	Day after	96.1	98.2	66.8	60.7	0.29	0.31	0	0	0.172	3.417
STD_B	Day before	94.7	95.5	96.2	95.5	0	0	0.06	0.11	7.778	16.02
	Event day	91.3	68.7	84.5	66.1	0.58	0.91	3.51	18.8	9.413	25.33
	Day after	90.4	49.4	85.1	42.9	1.48	7.56	9.47	50.7	8.547	15.78
STD_C	Day before	96.1	97.4	97.1	97.4	0	0	0	0	1.668	4.244
	Event day	59.7	23.2	57.3	44.4	25.2	32.8	28.4	58.2	4.284	19.61
	Day after	88.2	61.4	75.3	43.9	6.02	12.3	12.6	32.1	1.794	10.63

day, which coincided with the time that the spectators left the stadium). It also affected all the daily values for the next day of the event, by significantly increasing them. The day before the event is not affected, so it sets the normal values with which the congestive ones should be compared.

The cell STD_A covers mostly the area of the stadium itself, as understood by the values of the indicators: Actual traffic is present only at event hours; CSSR degrades to 32% that day and nearly disappears (12%) at the event's peak (busy hour of the event day). BR and HOSR present the same behavior. After the event, all indicators return to normal values, except the HOSR, thus making it clear that the crowd is moving to neighboring cells (actually leaving the stadium) and that STD_A covers mostly the stadium itself.

The cell STD_B mostly covers neighboring areas to the stadium, as increased blocking and call failures were observed when the crowd moved from the stadium (measured in the day after values). The highest values of blocking and failures appear the day after the event, thus making it obvious that STD_B covers areas out-side the stadium mostly. Even in that case, the situation is clearly better than the one encountered in STD_A.

Finally, cell STD_C is between the two situations described above. It covers part of the stadium area and part of the areas around it. In that way, it presents values below average not only on the day of the event (especially during the busy hour, val-ues reach the lowest points), but also in the day after the event (where HOSR reaches its lowest level).

It is very interesting to estimate what is the loss for the operator, during such congestion situations, apart from the decreased user satisfaction. Estimating the extra traffic produced by the event in Erlang, an amount of around 250 additional Erlang is requested only in the busy hour of the event. Supposing that the event lasted at least four hours (but traffic was lower for the nonbusy hours), at least 600 additional Erlang were requested during the event (optimistic estimation). The measurements show that the three cells served about 200 Erlang at the most (also an optimistic estimation). The result is that 400 Erlang were not served and thus were lost for the operator. The traffic that could not be accommodated is easily translated into profit loss, taking into account the cost of a call minute. It is obvious that cellu-lar operators lose a huge amount of profit during such sport events due to the not-served calls, still not considering user satisfaction, which is difficult to estimate but can lead to various problems for the operator. Later in the book user satisfaction will be considered in proposed business models for cellular operators.

2.2.3.6 Handover Performance

Handover can be caused by many different reasons. Each mobile terminal attempts to use the radio channel that will provide the best connection quality [i.e., the best carrier-to-interference ratio (C/I)]. Cochannel interference is a factor that cannot be avoided because of multiple use of the same time and frequency channels due to existing cell layouts, and consequently quality can be low (i.e., a high bit-error ratio) despite a high signal level. A usual cause of interference to other mobile stations may be the connection of a mobile terminal to the base stations, even if it is a high-quality one. The interference can be minimized if the interfered station changes to a different radio channel. It is also possible for mobile users to have the same good reception quality from more than one cell. The service quality of the network can then be optimized if mobile users are equally distributed over the available cells.

A connection is continuously being measured and evaluated by the respective base station and the mobile. Handover execution is based upon this evaluation. The decision algorithm mainly contributes to the spectral efficiency of the radio network and the service quality as seen by the mobile subscriber. A mobile subscriber leaving the coverage area of a base station must receive coverage from a neighboring base station to keep a connection intact. Connection cut off or "call drop" is not acceptable to the mobile user during conversation. When the traffic volume of a cell periodically reaches too high a level or when neighboring cells are being underutilized, the network itself can also initiate handovers. The main reason for a handover request is the uplink/downlink level, while power budget is the second most important reason for handover initiations. Other reasons are downlink quality, umbrella-cell handover, interference, directed retry, and OMC shortcomings.

The HOSR is strongly linked with the location of the two cells. Obviously the BSC-controlled handovers are easier to be handle; therefore, the success rate is higher. Another conclusion is that the performance of MSC-controlled handovers is significantly worse, since more network elements participate in this procedure, and therefore, the failure can occur in more places.

2.2.3.7   Other Types of Congestion

This part analyzes other types of channel blocking during a congestion situation, where most interfaces are overloaded, something that can influence the overall performance of the network.

2.2.3.7.1   Paging

This part of the study is focused on the paging shortcoming during congestion. Paging is most of the time not considered as a major problem, since it is not easily detected by the subscribers. Obviously, only mobile-terminated calls (MTC) and the terminated short message service (SMS) are considered. When a call request reaches the MSC, it is forwarded to the BSC that serves the LA of the user. In most of the cases, a BSC serves one LA. In any other case, the MSC will send this command to all the BSC under the same LA. Subsequently, the BSC distributes the paging request to all BTSs belonging to that LA. If traffic congestion is detected, the number of paging messages is dramatically increased, and this can cause problems. Once again the bottleneck is the A_{bis} interface that carries all the messages between the BTS and the BSC. An average value considering the length of a paging message is that a 16-Kbps A_{bis} link can handle around 100,000 paging requests per hour.

2.2.3.7.2   Congestion in the RACH

The shortcomings that are related to RACH are very difficult to measure, since the call request is often blocked at this early stage. On the other hand overload of BSC against high load in the RACH can be detected, and blocking itself does not take place in the RACH, but in the capacity of the BSC to answer all of the RACH retrieved by all of the BTS.

For that purpose we only consider mobile-originated calls (MOCs). Due to traffic conditions of the area, the number of RACHs listened to by the BTSs can be quite large, because the originated SMS attempts can be added to the mobile originated calls. The channel request messages sent to the BSC are equal to the number of RACHs received by the BTS. There is a protection in the BSC against this high RACH load that enables the processor to delete or omit some RACHs. All channel request messages received by the BSC should be answered with the messages that are subtracted from it. In any other case the BSC would have omitted the RACH due to traffic overload.

Note that when the BSC sends the immediate assignment rejection to the MS, one of the reasons could be that the A_{bis} interface has no internal resources to handle the request. In this case, inside the message there will be a message "wait indication," which defines the wait period for the MS.

2.2.3.7.3   Congestion in the AGCH

Once the network has reserved one SDCCH for the MS, the BTS has to send back information noting which channel it has reserved for it, the timing advance, and whether it uses frequency hopping. All this information is sent in the message "immediate assignment." It is directed in the logical channel AGCH. Hence another possible source of blocking in high traffic conditions could be AGCH blocking.

To calculate AGCH blocking, BSC measurements are used. The BSC sends to the BTS the immediate assignment or immediate assignment rejected commands. In the BTS there is one AGCH buffer that permits the BTS to store some of those commands coming from the BSC. In case the buffer is full, the BTS will respond with a delete indication. Thus, the ratio of delete indication to the sum of those commands describes the blocking of AGCH.

2.2.3.8   Summarizing Congestion Diagrams

Before any attempt is made to characterize the traffic load scenarios, based on a piece of predictable information, such as occurrence time, congestion localization, duration of the event, and amount of traffic observed during the congestion, it must be clear how the information from the DW should be interpreted to summarize congestion diagrams. Therefore, utilization and blocking of certain channels should be corresponded to performance indicators such as BR, success rate, and throughput of the respective channels.

Quantitatively corresponding the abovementioned indicators is a difficult task with doubtful results that would add overhead to the whole process. Thus, a qualitative relation is attempted instead. For that reason, three charts are generated, depicting the network's reaction in congestion situations for the utilization of the channels. The basic aspects of the chart are explained in Figure 2.26.

Low congestion means that only a few BSCs are affected and probably is geographically limited; medium congestion means that a respective percent of BSCs participate in the crisis; and high congestion means that almost the entire network is affected. The above classification of situations is also based on the results on the network level.

At first, traffic can also be considered as the throughput of the network. As seen in the diagram, it increases beyond the threshold of the low congestion situation but

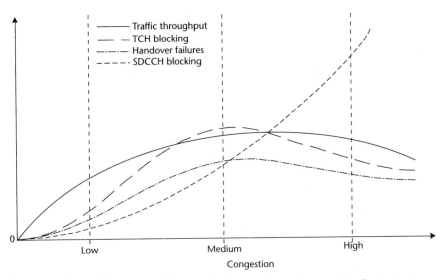

Figure 2.26   Network's behavior in congestion situations for performance indicators.

reaches a limit beyond the threshold of the medium congestion conditions. This limit represents exactly how much traffic the network can handle (or the maximum throughput the network can achieve). After the threshold of the high congestion, traffic is slightly degrading, as the conditions are worsening and the signaling resources required for a service (mostly SDCCH) are dramatically decreased.

As far as SDCCH blocking is concerned, it is clearly marked on the diagram that it is increasing proportionally to the augmentation of the congestion level. This is normal, because the demand for establishment resources increases, while resources are static and limited. In that way, blocking is increased proportionally to the demand. The importance of SDCCH blocking in the whole study is that all procedures are SDCCH-dependent.

TCH blocking also increases as traffic (or demand) increases, beyond the low congestion threshold, and reaches a peak at medium congestion, according to the traffic. As high congestion situations approach, TCH blocking is clearly reduced. This is also SDCCH-dependent, and it occurs because the network has run out of signaling resources (SDCCH mostly), traffic is reduced, and there are TCHs available. It must be noted that, in cellular networks, TCHs outnumber SDCCHs in every single TRX.

Finally, *handover failure rate* follows the curve of TCH blocking (the same applies for call setup failure rate, which is 1-CSSR). It is also SDCCH-dependent, reaches its peak when TCH blocking is at its highest point, and diminishes at high congestion conditions, because services cannot be established at this point.

According to the above facts, the utilization diagrams have the same behavior. It is very important, at this exact point of the analysis, to clarify that the following charts slightly precede the one above in terms of time and network monitoring. In other words, monitoring a channel, one will first perceive a change in utilization, followed by a change in blocking/success rates. A graphical explanation of this situation is shown in Figure 2.27.

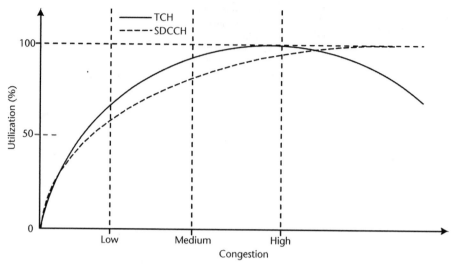

Figure 2.27   Network's reaction in congestion situations for utilization of SDCCH and TCH.

As shown above, TCH suffers the most from the congestion situation, followed by the SDCCH. The utilization of the two channels is similar in both low and medium congestion situations. When SDCCH gets highly congested, TCH utilization degrades, as there cannot be any more TCH assignments (lack of setup resources). If the traffic load is even higher, the situation worsens; RACH and PCH overcome the highest level of utilization, and SDCCH utilization starts to degrade too, as call setup attempts are now not perceived at all by the network.

In Figure 2.28, the behavior of RACH, PCH, and AGCH utilization is depicted. These channels follow a smoother increase of their utilization, as there are usually enough resources to handle the attempts made. AGCH, especially, rarely encounters blocking or utilization problems. The most common phenomenon, though, as far as

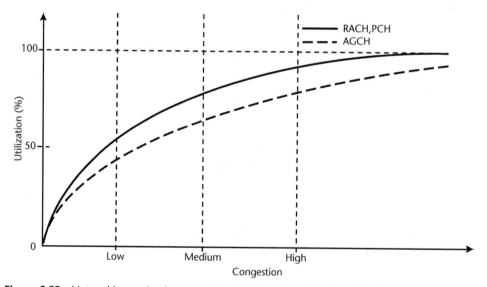

Figure 2.28   Network's reaction in congestion situations for utilization of RACH, PCH, and AGCH.

RACH and PCH are concerned, is that problems are encountered in the A_{bis} interface (air interface resources are usually adequate to handle most situations) [26], causing congestion situations. Usually, all three channels (PCH, RACH, AGCH) remain at medium or low congestion levels.

Combining the information from the three aforementioned charts, it is feasible to relate utilization and BR, reported by the monitoring tools, to congestion and, therefore, to characterize the scenarios. Utilization is the key indicator that will characterize a congestive scenario, as it is more representative of what is really happening and more independent from the situation itself than blocking. On the other side, blocking will provide additional information and, thus, extra help in the characterization of the scenario. BR depends on certain network parameters, such as radio coverage of the selected area, overlapping of the cells, and the use of directed retry [27, 28]. This "drawback" can be used in favor of the method, as it will offer extra characteristics to the situation. For instance, in a congestion situation where utilization reaches high values, it is very possible for blocking to remain at a low level because of good cell overlapping and proper use of the directed retry feature. Another possibility is that, while utilization decreases to very low values giving the impression that everything works fine, BR reaches very high values, raising in that way the alarm that channels are not used because of congestion. The use of utilization in combination with channel blocking will be obvious in the construction of the characterization rule for each scenario.

A further significant point of this analysis is that most of the events are scaling events, meaning that the increase will pass through all the congestion levels, on the road to the highest peak, and that must be taken into consideration on the selection of the appropriate scenario.

2.3 Traffic Modeling

Traffic engineering and modeling in communication networks is very important since it allows us to better understand the network's behavior under various conditions and thus to better organize its operational deployment and usage—or in other words to match the demand for traffic with what the network offers. A the same time traffic engineering and modeling in mobile communication networks apart from being important it is also a very complex and difficult procedure because of the nature of wireless communications. Thus it does not come as a surprise that ITU-T has developed a complete set of recommendations—the so-called E.750 series—to tackle this issue [29, 30].

Traffic engineering in mobile networks is important, as has been shown in the previous sections of this chapter, both for traffic channels and signaling channels and has distinct differences from the methods corresponding to fixed networks. In particular in mobile networks we have the following facts:

- Bandwidth is limited and the probability for the occurrence of transmission errors is rather high.
- A small number of active users can absorb most of the traffic capacity available in a cell.

- The main "bottleneck" is the radio interface at the edge of network (i.e., the availability of physical channels).

User mobility in particular affects signaling traffic, and for that reason models that predict the handover rate are required, usually called mobility models, to accurately dimension the signaling traffic that is generated from this process. Accurate mobility modeling is one of the big challenges in mobile networks. A big number of mobility models has been devised by the research community; a categorization of these is given in [31, 32].

According to recommendation E.751 of ITU-T [33] there exist two interfaces at which traffic demand has to be modeled to achieve more efficient traffic dimensioning and control. The first one corresponds to the radio interface while the other one is related with the modeling of the traffic that originates from the mobiles and is subsequently inserted in the fixed network.

The main input that we have to estimate the traffic demand is the size of the population that is going to be served by the cellular network, together with the year-on-year user penetration forecast and the average traffic intensity per subscriber. To be more accurate we have to partition the users into specific geographic areas, since other demands and behavior have the people in rural areas and other in urban ones. In recommendation E.760 of ITU-T [34] specific methodologies for estimating traffic demand in mobile cellular networks in general and in relation to the radio interface are introduced.

2.3.1 Modeling Approach

To effectively model traffic in mobile cellular networks the usual procedure is to develop three different models: a topology model, a call model, and a mobility model [32].

The topology model describes the geographical area in which the cellular network is deployed and consists of cells and the associated base stations and a dynamic set of mobile hosts.

The call model is responsible for describing the calling behavior of the mobile user with a set of representative parameters and their associated distributions. Such parameters include call duration and call initiation interarrival time in a specific cell. The call model may include also transmission and reception of short messages as well as other types of service usage like WAP and GPRS. In general, calls are generated according to a Poisson process, which is automatically satisfied when the user population is large enough and when users place their calls independently.

Finally the mobility model describes the effects of a mobile user's movement on traffic and signaling channels. The basic procedures that are included in the mobility model are location update and handover.

Apart from the models accurate reference loads are important to acquire a correct insight into the network's behavior. Reference loads can be extracted either from measurements of operational cellular networks, or they can be formulated using analytical methods. One important difference in mobile networks in relation to the fixed ones is that a big portion of signaling traffic is independent of call attempts and connection establishments (e.g., due to periodical location updates) [35].

2.4   Traffic-Load Scenarios

Section 2.2 presented the analysis of DW data from a cellular operator, focusing on the behavior of the network during congestion situations. The conclusion of the studies is that logical channel utilizations combined with BRs can characterize in a unique way each traffic-load scenario. In the following sections a number of traffic-load scenarios are presented and characterized, based on the outcome of the statistical evaluation [36].

2.4.1   Busy Hours

In telecommunication systems a BH is defined, as the sliding 60-minute period during which the maximum total traffic load in a given 24-hour period occurs. Since BH is a periodical situation, the time occurrence, traffic overload, and location of the event are predictable with a good confidence degree. In particular, the parameters presented in the following paragraphs are estimated based on the statistical analysis of the observed events.

The cellular operator can either define the BH of the day for the whole network, averaging all available data, or even define a BH for each MSC or BSC area. The starting time of the congestion event, as well as its duration, are statistically predictable. Depending on the area and time of the year, three different BHs can be defined. First, in a metropolitan area, during winter, BH occurs around the 12- to 1-P.M. time window. In the rest of the country, during winter, BH occurs in the 8- to 9-P.M. time window. Finally, during summer, BH occurs in the 8- to 9-P.M. time window almost everywhere. It should be mentioned that there are deviations from these periods that are related with the mentality and behavior of the subscribers from different countries.

The channels subject to congestion are mainly the TCH and the SDCCH. TCH utilization can overcome 80% during the BH, while SDCCH utilization reaches values up to 70%. The channel that first gets congested is the TCH, while SDCCH follows. It should be noted that in certain cells of a metropolitan area, utilization can reach 100%. As far as the other channels are concerned, PCH, RACH, and AGCH remain in the low congestion utilization range. BRs in this scenario are 1 to 10% for the TCH and 1 to 5% for other channels. Of course, as mentioned before, blocking depends highly on the area, radio coverage, and other network parameters, and these values are averaged estimations to help the identification process.

The cells taking part in the BH events are also predictable. A list of such cells can be provided, depending on the specific geographic area involved in the BHs. As mentioned above, the cells mostly affected are the ones serving urban areas, especially if they are highly populated. This fact means that only a few cells in the entire network encounter high congestion on BHs.

2.4.2   Holiday Resorts

We consider holiday resorts the places where a considerable number of people spend their holidays (e.g., the seaside, islands, mountains, and lakes). Even if the number of tourists in a given location can vary over the years in a rather unpredictable way, the traffic overload can be estimated fairly based on statistical analysis of the data collected in previous years.

In very simple words, what can be expected during the tourist period in a specific location is a "scaling" effect, which multiplies the time-dependent utilization factor of the network resources proportionally to the amount of users in the area affected by the tourist flow. Thus, areas in which the local BHs do not generate any network congestion during nontourist periods may be daily affected by an excessive degradation of network performances in those periods, during which the number of wireless users is multiplied by a factor of two, five, and even ten.

Congestion characteristics of an event belonging to the traffic load scenario should be quite similar to that of the BH presented above. Slight differences are observed in the utilization factors of certain channels (which are slightly increased), in the starting times of the traffic peaks, and the scenario's predicted duration. The cells mostly affected in this scenario are the ones serving tourist destinations that are easily located.

The channels subject to congestion are mainly TCH and SDCCH. Both TCH and SDCCH utilization can overcome 80% during the BH of this traffic-load scenario. The channel that first becomes congested is the TCH, while SDCCH follows. As far as the other channels are concerned, both PCH and RACH climb to the medium congestion utilization factor, while AGCH rests at the lowest level of congestion. BRs in this scenario remain at 1 to 10% for the TCH and at 1 to 5% for the other channels. Of course, as mentioned before, blocking depends highly on the area, radio coverage, and other network parameters, and these values are averaged estimations to help the identification process.

Both the starting time of the congestion event, as well as its duration, are statistically predictable. Regardless of which day of the week is considered or the area affected, the traffic peak is most of the times reached between 8 P.M. and 9 P.M. as people tend to move their acting hours later than usual.

The cells involved in this scenario event are predictable. A list of such cells can be provided, depending on the specific geographic area where the tourist movements are taking place.

2.4.3 Bank Holidays

We consider as *bank holidays* those public holidays such as New Year's day, Easter, Christmas, and federal holidays. In these particular days of the year people enjoy contacting relatives and friends, thereby causing a huge growth of the traffic load in a relatively short period of time. Such traffic overload periods (e.g., the minutes around midnight on New Year's Eve) are fairly easy to predict from previous statistical data.

All channels are affected in such a scenario, but AGCH reaches first the medium utilization factor and then is increased to the highest level of utilization. Subsequently TCH becomes congested, followed by the SDCCH. When SDCCH gets highly congested, TCH utilization degrades, as there cannot be any more TCH assignments (lack of setup resources). If the traffic load is even higher, the situation worsens, RACH and PCH overcome the highest level of utilization, and the user has practically no cell phone. TCH BR usually overcomes the 10% threshold, while

SDCCH, RACH, and PCH also reach high values, way over the 5% threshold. Only AGCH remains in a medium blocking situation.

Both the starting time of the congestion event, as well as its duration, are statistically predictable. As observed during the statistical analysis, the time of the event depends on the bank holiday. For instance, the New Year's Eve congestion takes place between 11 P.M. and 1 A.M., while for other bank holidays, congestion is observed during the day. The most important aspect of this scenario is that the date is also predictable. Yet, the most predictable case is causing the greatest problem to the network.

The cells involved in the bank holiday congestion events are predictable. Usually, this scenario applies to the whole network's topology. The event is expected to affect the network level, and only a small number of cells will behave well.

2.4.4   Sporting Events

During *sporting events* a large number of persons are concentrated in a limited location. The crowded cells pass from the average concentration of users to a much higher number of users than it can accommodate for a period of about 2 to 3 hours, depending on the duration of the sport event. The time and location of the event are known in advance, and the traffic overload can be estimated on the basis of data that has been previously collected during similar events.

All channels are affected in such a scenario, especially the AGCH that reaches first the medium utilization factor and this is then increased to the highest level of utilization. In addition, TCH is strongly affected by the congestion situation, followed by the SDCCH. When SDCCH gets highly congested, TCH utilization degrades, as there cannot be any more TCH assignments (lack of call setup resources). TCH BR usually overcomes the 10% threshold, while SDCCH, RACH and PCH also reach high values over the 5% threshold. Only AGCH remains in medium blocking. If the traffic load is even higher, the situation worsens; RACH and PCH overcome the highest level of utilization, and the user has practically no cell phone. The latter usually happens in certain periods of the event, such as "dead" periods (e.g., half time of a match) or extremely "active" points (e.g., scoring of a goal and gaining a victory).

The starting time and date of the congestion event, as well as its duration, are known in advance for each particular event. In that way no time window can be defined, but it will be easily statistically predicted and identified, based on previous statistical data.

The cells involved in the sport event congestion scenario are predictable. Usually, this scenario applies to very few cells serving the location of the sport event and its surrounding area. At first, the congestion appears at the surrounding cells, as the crowd arrives at the stadium. Then congestion moves to the cells serving inside the stadium, and, finally, it is again redirected to the surrounding cells, as the crowd is leaving.

2.4.5   Cultural Events and Demonstrations

Cultural events like theatre performances, movies, concerts, religious events, as well as public demonstrations cause a concentration of people in specific locations. On those occasions a medium or large number of people use their mobile phones mostly to call friends and relatives. Since the time and location of the event are known in advance, the traffic overload is quite predictable on the basis of data previously collected in similar events.

Usually, the dimension of this event is slightly smaller than the one of a sporting event, as far as channel utilization is concerned. Both TCH and SDCCH reach the highest level of utilization, but the other channels remain at the medium utilization factor. TCH is the first network resource that is affected by the congestion situation, followed by the SDCCH. When SDCCH becomes highly congested, TCH utilization degrades, as there cannot be any more TCH assignments (lack of setup resources). TCH BR varies between 1% and 10%, while SDCCH BR is 1 to 5%. The other channels are in the range of low-level BRs.

The starting time and date of the congestion event, as well as its duration, are known in advance for each particular event. Thus, one cannot define any time window, but time windows are easily statistically predicted and identified, based on the statistical data.

The cells involved in the cultural events congestion scenario are predictable. Usually, this scenario applies to very few cells serving the location of the event and its surrounding area. At first, the congestion appears at the surrounding cells, as the crowd is coming to the location of the event. Then congestion moves to the cells serving inside the theater or cinema and, finally, is redirected to the surrounding cells, as the crowd is leaving.

Under this traffic-load scenario, demonstrations are also included, since the behavior of the network has many similarities in these cases. Demonstrations for political, social, and trade union reasons are potential congestion situations. The traffic overload depends mostly on the concentration of users and partially on the participants that communicate and exchange information about the demonstration progress. Since the time and location of the event are known in advance, the traffic overload is predictable to a certain extent if data has been collected in similar events.

2.4.6 Network Shortcomings

The *network shortcomings* can be classified to those that are planned (e.g., maintenance) and those that are unexpected. In case of a planned outage for repairing the network, the service in the corresponding cell(s) is either unavailable or limited. The traffic trend is the usual one, but because of the outage a certain number of users are redirected to the adjacent cells, causing a traffic overload in those areas. The location and duration of the outage are known in advance. The number of users that are redirected in the adjacent cells can usually be estimated on the basis of statistical data.

The channel that is mostly affected in this scenario is the SDCCH, especially if a location update is demanded for the users directed to the neighboring cells. Thus, it reaches the highest values of utilization. TCH can vary its utilization between the medium and high scale, while the other channels remain in medium (RACH/PCH) or low (AGCH) congestion scales. Blocking is assumed to be high in such a situation for all channels, except the AGCH, which reaches medium values. High blocking is mainly caused by the redirection of users to the neighboring cells, but highly depends on the characteristics of the area of the outage.

The starting time and date of the planned outage, as well as its duration, are known in advance, but for each event in particular. In that way it is impossible to define any time window, but it is easy to predict and identify one based on previous statistical data.

The cells involved in the planned outages congestion scenario are known in advance. This scenario applies to few cells serving the surrounding area of the location of the outage. When the planned elements go down, all the traffic is redirected to the neighboring cells.

In case of an unexpected BTS or BTS link shortcoming, users are redirected to the nearest BTSs, where the traffic overload situation takes place consequently. The location, time, and duration of the shortcoming is not known in advance. The traffic overload can be estimated to a certain extent on the basis of the average number of users of the cell corresponding to the damaged BTS.

The starting time and date of a BTS shortcoming, as well as its duration, are unpredictable. In that way no time window can be defined; however this can only be roughly estimated based on previous statistical data. The cells involved in the BTS shortcoming congestion scenario are known in advance. This scenario applies to a few cells serving the surrounding area of the BTS that is out of order.

In case of an unexpected BSC or a BSC link shortcoming, some areas inside the BSC coverage could remain isolated. In a BSC with a star topology, users located at the edges of the BSC coverage area are redirected to the nearest BSCs; otherwise their calls are not satisfied. The location, time, and duration of the shortcoming are not known in advance. The traffic overload can be estimated to a certain extent on the basis of the average number of users at the edges of the area related to the damaged BSC.

The starting time and date of a BSC shortcoming, as well as its duration, are unpredictable. In that way no time window can be defined; however this can be again roughly estimated based on previous statistical data. The cells involved in the BSC shortcoming congestion scenario are known in advance. This scenario applies to the cells serving the surrounding area of the BSC out of order, meaning the cells on the border of the respective BSC. When the respective BSC goes down, all the traffic is redirected to the neighboring cells.

2.4.7 Catastrophes

After small, medium, and large *catastrophes* like earthquakes, floods, volcanic eruptions, and ecological disasters, an increase of the traffic load is usually observed, due to emergency calls or simply calls to inquire after the safety and health of relatives and friends. Also damages in the network could worsen the traffic status. The traffic is not predictable. A statistical estimation can be made on the basis of the data collected from previous similar events.

All channels are affected in such a scenario. All channels increase to the highest level of utilization. TCH is the first channel that is affected from the congestion situation, followed by the SDCCH. When SDCCH gets highly congested, TCH utilization degrades, as there cannot be any more TCH assignments (lack of call setup resources). If the traffic load is even higher, the situation worsens, RACH and PCH overcome the highest level of utilization, and the user has practically no cell phone.

Blocking in such a case will be high on all channels. These are the cases most likely for the network to collapse, if the catastrophe is extended.

No starting time or date can be predicted, as the event is totally unpredictable. As far as the duration is concerned, it depends on the dimension of the catastrophe and can be predicted only by using previous statistical data. Thus, no time window can be specified.

The cells involved in the catastrophic congestion events are not predictable. Usually, this scenario applies to the area where the catastrophe has taken place. Depending on the dimension of the catastrophe, the affected cells can be easily defined, as they will not behave normally.

During catastrophes, the scenario can also be characterized as massive. The *massive congestion scenario* is a very particular one, since it happens only when the request of radio resources reaches extremely high intensity. This is the case, for instance, when an outage causes some parts of the cellular network to be uncovered for a short period of time, and when the network comes up again, there is a huge peak of signaling traffic generated by the mobiles that regain their access to the network.

Under these circumstances, the extremely high rate of requests may overload some channels (especially signaling) up to the point that the resource utilization actually goes down because of the bottleneck effects. This scenario can be uncovered by looking at the BRs and at the clear codes for the calls. Therefore, by combining the information provided above, we come up with the following characterization rule to capture the massive congestion scenario.

2.4.8 Accidents

The *accident scenario* mainly refers to such events as road accidents, queues, and acts of terrorism. When an accident happens, involved people make not only emergency calls, but also calls to relatives and friends. At first the event is delimited in a restricted area, but can quickly spread out in adjacent areas—for example, a road accident can block the traffic for many hours, causing queues and slow traveling. The traffic is not predictable. Some very rough statistical estimations can be made on the basis of the data collected from previous similar events.

The channels subject to congestion are mainly the TCH and the SDCCH. Both TCH and SDCCH utilization can overcome 80% in such a case. The channel that first gets congested is the TCH, then the SDCCH. As far as the other channels are concerned, both PCH and RACH climb on the medium congestion utilization factor, while AGCH rests at the lowest level of congestion. TCH BR is expected to overcome the 10% threshold, while SDCCH, PCH, and RACH BRs are estimated in the medium blocking window. AGCH probably will not encounter any blocking problems.

No starting time or date can be predicted, as the event is totally unpredictable. As far as the duration is concerned, it depends on the dimension of the accident and this can be roughly predicted using previous statistical data. Thus, no time window can be specified.

The cells involved in the accident congestion events are not predictable. Usually, this scenario applies to the area where the accident has taken place. Depending on the dimensions of the accident, the affected cells can be easily defined, as they will present an extra traffic load.

2.5   Resource Management Techniques and Guidelines for Implementation

This section provides a detailed description of radio resource management techniques and mechanisms that can be applied in a cellular system. Cellular networks in general, are managed in both a distributed and centralized manner. The proposed management techniques are based on the concept of constantly real-time network monitoring and intelligent decision-making. Therefore, most of the techniques can be seen as a dynamic reconfiguration of network parameters to bring the network to a stable position. This approach tackles congestion shortcomings by an automatic-control, closed-loop system that reconfigures the networks needed by applying a set of techniques [37, 38].

The real-time monitoring is based on distributed evaluation of reports generated by the network (mostly by the MSCs), during call set up, call termination, as well as other procedures. The evaluation of reports can be used for the measurement of the characteristic KPIs of the network, like the ones presented in the previous sections. For each KPI and each cell, thresholds can be defined, so that it will be possible to generate alarms whenever congestion is detected. These alarms can be forwarded to a centralized resource management unit. This unit is responsible for determining the size of the congestion, by requesting information from the adjacent cells and assigning the most appropriate traffic load scenario to a cluster of cells affected by congestion. Subsequently, historical information about the techniques' effectiveness can be utilized to select the appropriate resource methods to be applied in the cluster of cells, as well as for the fine-tuning of each technique. After congestion is tackled, the system can restore the default values of the parameters and return to the initial state.

2.5.1   Half-Rate and Full-Rate Usage

Cellular operators can make use of full-rate TCH (TCH/FR) and half-rate TCH (TCH/HR) by modifying the channel rate. With HR coding the operator can maximize the spectrum efficiency, comparing FR traffic channels with HR coding. On one hand, the use of TCH/HR results in doubling the amount of TCH resources that is available for traffic. On the other hand, voice quality is obviously degrading, but it is a trade-off decision for the operator, which must determine whether the change will increase the overall benefit.

FR speech and data is coded and transferred by using 16-Kbps channels in the BSC. With HR coding, the transmission requires 8-Kbps channels instead. The low-rate data services can be employed with HR traffic channels. In the HR the MS uses only the 26 frames from the 51 frames that are used in the FR and enhanced FR (EFR), since the network resources are doubled.

The FR/HR selection feature can be implemented if the network can support dual rate requests and when both channel rates are acceptable. Most of the network manufacturers allow for the use of this feature. For requests that determine a single TCH rate a channel of the respective type is allocated regardless of the traffic channel load in the BTS.

2.5.1.1 Resource Management Technique Parameters

This resource management technique can be controlled in most vendors by modifying two parameters. The first is the lower limit for free TCH/FR resources, and the second one the upper limit for free TCH/FR resources. These parameters can either be defined on the BSC or BTS level. The parameters can be given as relative amounts of the available TCH/FR resources.

When the number of the free TCH/FRs is below the threshold for the lower limit for free TCH/ FRs, the TCH/FR slots are split into TCH/HRs. In the same way, when the number of free TCH/FRs is above the upper limit for free TCH/FRs, all the allocated free TCH/HRs are transformed back to TCH/FR. Figure 2.29, shows an example of upper and lower limits for free TCH/FRs.

TCH/HR channels can be used both for voice and data calls. The air interface has to support TCH/HR allocation to implement this feature to the network.

This resource management method can be applied immediately upon request with very good results in congestion situations where there is shortage of TCHs. Whenever this technique is applied, the problem that a number of MS request only TCH/FR allocation, can be solved but the method's effectiveness degrades.

The disadvantage for the traffic channels with HR implementation is that they provide a lower quality of sound compared to the one of FR coding. The implementation of this technique is not suitable for high congestion situations as the main problem is located in the SDCCH channels. Only in the case of TCH congestion can the TCH/HR allocation can be implemented.

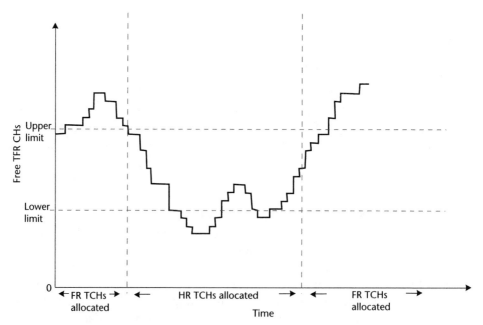

Figure 2.29 Limits for free TCH/FRs.

2.5.1.2   Expected Results

Let us assume there is a BTS with 23 TCHs and that the value for the lower limit is 18%; that means the threshold is the 19 full-rate TCH used channels. In addition the upper limit is 35%, which points to a threshold of 15 full-rate TCH used channels.

The maximum blocking probability that a user will face up is 6.6% for full-rate channel allocation. The value has been calculated from the Erlang B-table, as the available traffic channels are 23 and the offered traffic 19 Erlang, which is the outcome from 19 used traffic channels within an hour. Therefore, the available channels will be four TCH/FRs (state 1), as shown in Figure 2.30.

When the feature is initiated the available TCH/FR channels will split to TCH/HR, thus to eight TCH/HRs and the total traffic channels from 23 will become 27 (19 TCH/FRs + 8 TCH/HRs). Calculating via the Erlang B-table, the new blocking probability drops to 1.8%, (state 2), as shown in Figure 2.30.

On the other hand, concerning the upper limit, which has been set to 35% (15 TCH/FRd), the calculated BR from 15 Erlang (15 TCH/FR occupied with an hour) and total of 27 traffic channels (8 TCH/HR + 19 TCH/FR), is 0.2%, (state 3) as shown in Figure 2.30. When the available channels exceed the threshold of 15 available TCH/FR then HR channels are converted back to full rate channels with total traffic channels equals to 23 (TCH/FR). Therefore the blocking probability becomes 1%, (state 4) as shown in Figure 2.30. In Table 2.7, the blocking probability according to the number of TCHs is presented for the two configurations.

2.5.2   Forced Handover

The handover procedure is the mechanism that transfers an ongoing call from one cell to another as a user moves through the coverage area of a cellular system. The number of cell boundary crossings increases, because smaller cells are deployed to

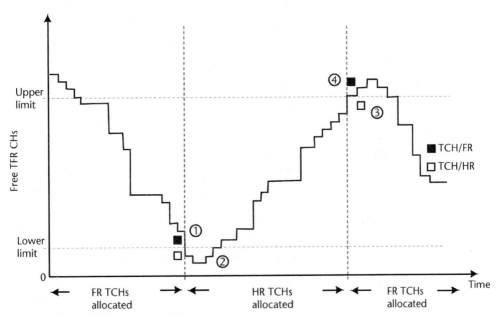

Figure 2.30   TCH H/R configuration.

Table 2.7   Blocking Probability According to the Number of TCHS and the Configuration

	Lower Limit Free TCH/FRs = 18% Upper Limit Free TCH/FRs = 35%		Lower Limit Free TCH/FRs = 35% Upper Limit Free TCH/FRs = 68%	
	Blocking Probability	Total TCHs	Blocking Probability	Total TCHs
State 1	6.6%	23	1.0%	23
State 2	1.8%	27	0.0%	31
State 3	0.2%	27	0.0%	31
State 4	1.0%	23	0.0%	23

meet the demand for increased capacity. If we minimize the expected number of handovers the switching load minimizes as well, because each handover requires network resources to reroute the call to the new base station. In GSM, measurement reports, which are transmitted periodically from MS to BS on the SACCH that is assigned to each communication, are available for each connection. The repetition duration of the SACCH produces a fixed time grid of 480 ms in which the measurement reports occur. Obviously the handover procedure considers a set of parameters in such a way to avoid shortcomings. On the other hand, it is not possible to have a safe handover execution in some cases, since the location and direction of the user as well as the area characteristics are not known. In addition the execution of the process takes place after a period of measurement reports, and the handover procedure is triggered again.

During congestion situations the use of forced handover applies to a management technique that aims to move the mobile stations from one serving cell to another to empty a TRX or even a whole cell. Before the radio network recovery management initiates a forced handover, the decision should be based on the measurement results reported by the MS/BTS and on the various parameters set for each cell, as in a normal handover process.

On the other hand, the forced handover technique will probably cause an extra load in signaling on the A_{bis} interfaces of the relevant BSC. Generally the handover procedure increases the DCR, which is a very important performance indicator.

2.5.2.1   Resource Management Technique Parameters

The forced handover can be applied to the whole cell, which means that all ongoing calls will be transferred to adjacent cells, if there is cell overlapping and adequate resources. Therefore, the major requirement is that a monitoring system measures the utilization of all cells in the area and when congestion is detected, all calls will be forced to handover to the adjacent cells. Recent implementations in the BSC allow partial execution of forced handover, which means that only calls occupying specific slots will be forced to handover, and this is influenced by the selection of the operator. This approach is more effective since it limits the possibility of drop call, due to shortage of radio resources in the adjacent cells.

The benefit after the implementation of this method is that the specific cell, which faces up congestion, will be decongested by the forced handover of some users, depending on their priority [39]. Thus the network will work more efficiently. However, in the case of a wide congestion problem including more than one cell, the forced handover feature is not implemented because it will cause problems even to other interfaces, like A_{bis}.

2.5.2.2   Expected Results

This technique is expected to have a positive impact on network performance, since users can be served by a neighbor BTS that is less congested compared to the one to which the user is assigned.

2.5.3   Dynamic Allocation of Signaling Resources

The SDCCH is always used when a traffic channel has not been assigned and is allocated to a mobile station only as long as control information is being transmitted. It is used in case of a call set up, before allocating a traffic channel. Moreover, SDCCH signaling is used also for SMS service, location area updates, as well as authentication. Under certain situations like demonstrations and places like airports, ports, and other hot spots, congestion occurs mainly in the SDCCH signaling resources. To overcome the arising problem of SDCCH congestion, extra SDCCH channels are needed. Despite the fact that the operator could add extra BTSs in a network, this will require more effort for network planning and optimization. On the other hand, the use of dynamic SDCCH maximizes the capacity of signaling resources dynamically by subtracting available TCH channels, which are converted and added to existing SDCCH channels.

The dynamic SDCCH feature enables the configuration of SDCCHs resources according to the actual SDCCH traffic situation of a cell. When the BTS needs larger SDCCH capacity than normally, idle TCH resources are configured for SDCCH. When the SDCCH congestion situation is over, the extra SDCCHs resources are configured back to TCH resources.

The operator has to configure the BTS with the minimum static SDCCH capacity to be sufficient to handle the normal SDCCH traffic, where the BCCH TRX must have an SDCCH channel. When the actual SDCCH congestion situation has started and the last free static SDCCH has been allocated, the dynamic SDCCH resource is allocated to keep the maximum TCH capacity available all the time.

2.5.3.1   Resource Management Technique Parameters

This feature does not have any specific parameters for extra tuning. It is a method that can only be enabled or disabled. One of the critical decisions is when to enable the dynamic allocation of SDCCH resources. On the one hand, if the detected traffic load scenario is one of those that are strongly linked with enormous signaling requests, then dynamic SDCCH allocation can help accommodate more traffic in the system, despite the fact that TCH resources will be given for signaling. On the other hand, in other congestion situations, where calls have a long duration and the arrival rate is constant and not high, the use of dynamic SDCCH allocation is useless and can even worsen the situation.

2.5.3.2   Advantages and Disadvantages

This feature adds value in traffic cases where signaling is the only transmission mode during the connection to the network. SMS traffic and location updating are counted among them. In some special locations such as airport and ports, the location updating can produce sudden short-time SDCCH traffic peaks, which can now be handled without having to configure extra permanent SDCCH capacity for the sake of safety only.

The dynamic SDCCH method can be configured both to TRXs of the extended and normal range areas of the cell. The definition of which TRX is used is based on which part of the cell area the MS access is received.

The disadvantage of this feature is that the dynamically allocated SDCCH cannot be allocated for SMS, location update (LU), or handover, but only for voice and data call set up. Moreover, when there are no available static SDCCH resources, the congestion continues to be almost the same for SMS, LU, and handover as dynamic SDCCH is activated, but only the voice calls and data calls take advantage of the feature.

2.5.3.3 Expected Results

Using a radio network simulator and running a simple scenario the results show that one cell with 23 TCH channels and eight SDCCH channels is congested as shown in Figures 2.31 and 2.32, where the blocking probability as well as the utilization of those resources (before applying the technique) are calculated [23].

Therefore, by applying the dynamic SDCCH technique, the congested cell is reformed and the resources are now 22 TCHs and 16 SDCCHs as shown in Figures 2.33 and 2.34, where the blocking probability is calculated as well as the utilization of those resources (after applying the technique).

As seen from Figures 2.33 and 2.34 the SDCCH utilization drops from 89% to 25% while the TCH utilization increases from 128% to 140%.

In the same way the following numerical example shows the advantages of dynamic SDCCH. Assuming that there is a BTS with 23 TCHs and eight SDCCHS, the following occur during a congested situation:

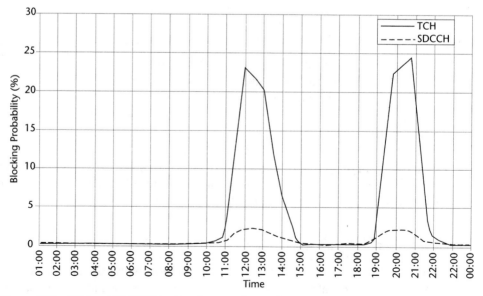

Figure 2.31   TCH and SDCCH blocking in normal scenario.

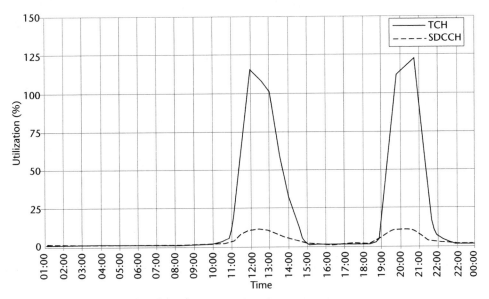

Figure 2.32   TCH and SDCCH utilization in normal scenario.

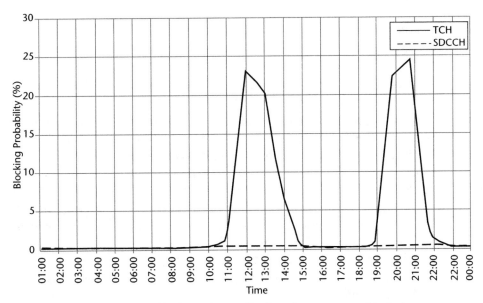

Figure 2.33   TCH and SDCCH blocking after application of a management technique.

- Blocking probability TCHs: 9.101%;
- Blocking probability SDCCHs: 28.147%;
- Blocking probability call set up: 34.68 %.

After applying one dynamic SDCCH/8, the following are true:

- Blocking probability TCHs: 11.26%;

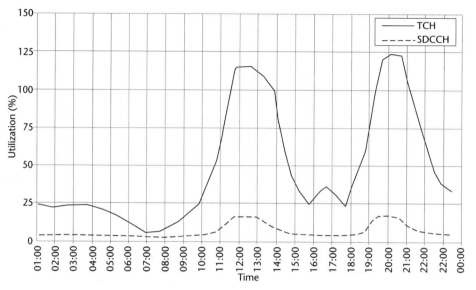

Figure 2.34   TCH and DCCH utilization after application of management technique.

- Blocking probability SDCCHs: 0.573%;
- Blocking probability call set up: 11.768%.

2.5.4   Prioritizing a TRX in TCH Allocation

It has been shown that in many cases the DCR arises due to radio interference. Congestion situations worsen due to the multiple use of a TRX within a cell. It has been observed that in a single cell, the users are assigned different TRXs, leaving available resources in each TRX. The use of all TRXs cause interference problem from cochannel existence in adjacent channels. The preferred solution is to group the assigned users to a minimum number of TRXs within a cell, hence in the network. Therefore, the TRX prioritization technique applies to solve this problem by concentrating the users in a minimum number of TRXs.

Allocation of traffic channels from specific preferred group of TRXs is reasonable if the TCHs of the group do not exceed interference thresholds. Calls that are assigned to a channel under heavy interference can be dropped and those channels can be allocated again for other calls, with the same consequences. Applying in TCH allocation the method of the minimum acceptable uplink C/N ratio, offers sufficient protection in these cases.

The normal practice in traffic channel allocation for a call is that when no quality requirements are stated, an attempt is made to allocate the TCH that has the least uplink interference. The TRX where the channel is going to be allocated is determined by the TCH resource situation in each TRX as well as by the rotation of resources in the TCH allocation. Similarly when TCHs fulfilling specific uplink quality requirements are searched for, none of the TRXs is given priority.

Sometimes it can be useful to favor the BCCH carrier for a call assignment. A reason for this is that the BCCH TRX transmits in all slots constantly the allocation of TCHs. The BCCH carrier, especially, does not increase network interference. The quality of the BCCH TRX channels is much better than that of the other TRXs since the BCCH carrier frequencies are not reused as frequently as the other carriers. Thus the quality of the TCH channel is better than that of the BCCH TRX and has a higher C/I ratio.

Priority setting between the BCCH TRX and other carriers in TCH allocation is a general GSM feature of the BSC. The TCH channel rate requirement when the resource is requested is determined in the TRX priority setting procedure. When either of the channel rates, FR or HR, is acceptable, then the TCH of the preferred channel rate is always allocated regardless of whether it is found from the priority TRX group or not. The channel rate preference can be set either initially by the MSC or by the BSC with its TCH allocation control parameters.

RF hopping can reduce the average interference experienced by he MS. RF hopping cannot be applied in the BCCH carrier, and therefore for quality reasons it is sometimes reasonable to assign a call priority to TRXs other than the BCCH carrier.

2.5.4.1   Resource Management Technique Parameters

The main parameter of this technique is to enable or not the prioritization in the TCH assignment. In addition, we can choose between a TCH from the BCCH TRX or from another TRX.

One advantage, arising from this feature, when the BCCH TRX has greater priority for the allocation of a TCH, is decreased interference on the network, due to fewer TRXs in use. Also it produces a better grouping of the TCHs to save gaps between slots for better performance into the multislot services, like high-speed circuit-switched data (HSCSD) and GPRS. Moreover, the blocking probability in multislot allocation as well as the number of dropped calls decreases.

The only disadvantage of this feature is that a BSC tries to gain access with less candidate channels available. When the feature is deactivated, the allocation of available channel starts among all the available slots with the successful candidate to become the one with the less interference.

2.5.4.2   Expected Results

The following example shows the benefits from the implementation of the TRX prioritization in the grade of service (GOS) in a multislot allocation. GOS is the probability of a call being blocked or delayed more than a specified interval, expressed as a decimal fraction. For the multislot allocation the GOS is given by:

$$GOS = 1 - \frac{\sum_{i=1}^{n} \left(\frac{\frac{T}{m}}{\frac{8}{m}} \right)}{n} \cdot \left(1 - GOS_{Singleslot} \right)$$

where n = number of TRXs, T = number of free TCHs in this TRX, and m = the number of allocated TCHs.

The $\binom{T}{m}$ calculates the groups of m TCHs that we may have from the amount of T TCHs. The effect of prioritization on a BTS with three TRXs is shown in Table 2.8.

Table 2.8 TRX Prioritization and Blocking Probability of Multislot Allocation

	TRX 1 (BCCH)	TRX 2	TRX3
Without Prioritization in the BCCH	4	2	1
With Prioritization in the BCCH	7	0	0

Without prioritization in the BCCH:

The $GOS_{\text{Singleslot}}$ calculated from Erlang B table for 8 Erlang offered traffic and 23 available channels is $GOS_{\text{Singleslot}} = 0\%$.

$$X = \left(\frac{\binom{3}{3}}{\binom{7}{3}} \right) + \left(\frac{\binom{6}{3}}{\binom{8}{3}} \right) + \left(\frac{\binom{7}{3}}{\binom{8}{3}} \right) = 0.356$$

$$GOS = 1 - \frac{0.356}{3} \cdot \left(1 - GOS_{\text{Singleslot}} \right) = 1 - 0.1187 \cdot (1 - 0) = 0.88$$

$$GOS_{\text{Multislot}} = 88\%$$

With prioritization in the BCCH:

$$X = \left(\frac{\binom{0}{3}}{\binom{7}{3}} \right) + \left(\frac{\binom{8}{3}}{\binom{8}{3}} \right) + \left(\frac{\binom{8}{3}}{\binom{8}{3}} \right) = 0.356$$

$$GOS = 1 - \frac{2}{3} \cdot \left(1 - GOS_{\text{Singleslot}} \right) = 1 - \frac{2}{3} \cdot (1 - 0) = 0.33$$

$$GOS_{\text{Multislot}} = 33\%$$

2.5.5 Dynamic Cell Resizing with the Use of C2 Values

The technique of cell selection is based on the idea that the mobile station should be within the cell offering the best coverage. In idle mode the mobile station has to find the best dominant cell in each geographic area. This process is called cell reselection and is based upon the comparison of C2 values from the cells that the mobile can receive. This method can be seen as "cell-breathing" in a GSM system for users in idle mode. The C2 value was known to the GSM as a method to control the reselection between different types of cells, like pico, micro, and macro. The dynamic cell resizing method uses this parameter dynamically to control the size of a cell, to avoid congestion under heavy telecommunication traffic situations, and especially to confirm that all the network recourses have been exhausted. Also this method leads to the best results when the congestion situation is predicted [40].

The C2 reselection information is broadcasted to the MS over the BCCH. The MS evaluates the C2 reselection criteria according to the given parameters. The C2 reselection criterion also includes the so-called penalty time (T_{pen}), which allows the cell reselection to occur slower than the C1 criterion evaluated by the MS. In this way undesirable reselection of a microcell in an environment where there is coverage both by a microcell and a macrocell is avoided.

This method does not affect other network elements, but only network planning. In situations, where the coverage area of a congested cell should be reduced, we might have a decrease of the C/I ratio of about 5 dB in the worst case. Also for cells that are located at the border of two different LAs, different settings are selected for cell reselection hysteresis to prevent useless LU.

2.5.5.1   Resource Management Technique Parameters

The idea is that the MS compares field strength levels coming from different cells within an area and selects the dominant one. The equation for the cell selection is configured as follows:

Cell selection in idle mode based on C1.

$$C1 = (Rx - \text{RxLevelMinAccess} - \text{MAX}(MSPTx - MSPmax), 0)$$

where:

- *Rx* is the signal level that the mobile terminal receives from the FCCH channel (lighthouse of the BTS);
- RxLevelMinAccess (*Rx* level minimum success) is the minimum receivable signal level, in which the MS may use the cell. This value is implemented from the operator and the values depend on the type and the role of the cell;
- *MSPTx* (mobile station transmission power) is the maximum allowable power that the BTS permits for the MS to gain access on the RACH channel;
- *MSMaxPwr* (mobile station maximum power) is the maximum RF power of the MS.

Thus, the MS takes into account the minimum access level to the cell and the maximum transmitting power allowed for the mobiles in each cell when starting a call.

Cell Reselection Criterion Based on C2.   With the introduction of multilayer cells: pico, micro, and macro, the need for priority and distribution of the traffic raised the subject of the usage of extra parameters for cell reselection. This led to the introduction of the penalty time (T_{pen}), which describes the time delay before the final comparison is made between two cells as well as the temporary offset, which, in turn, describes how much field strength could have been dropped during this penalty time. The cell reselection offset describes an offset to cell reselection. The C2-based cell reselection is calculated by the following equation:

$$C2 = C1 + \text{CellReselectionOffset} - \text{TemporaryOffset} \cdot H\left(T_{\text{pen}} - T\right)$$

when $T_{\text{pen}} \neq 640$,

$$C2 = C1 - \text{CellReselectionOffset}$$

when $T_{\text{pen}} = 640$

where $H(T_{pen} - T) = 1$ when $(T_{pen} - T) \geq 0$, $H(T_{pen} - T) = 0$ when $(T_{pen} - T) > 0$, and T the time when the mobile has received information from the new adjacent cell.

CellReselectionOffset.   To allocate the traffic load to another BTS, operators can change this parameter dynamically. This parameter has an effect on static or moving MSs.

TemporaryOffset.   This can be changed dynamically by the operator to allocate the traffic load to other BTS. This parameter only affects moving MSs.

To activate this technique, the network topology should be planned with overlapping areas between cells. On the other hand, when the size of the congested cell has become very small due to the very big values of CellReselectionOffset then the MSs that are located in the new border of the congested cell will face a C/I of less decibels than the one in planned for. So if the normal planning is very marginal (e.g., C/I = 11 dB), such a degradation in the C/I will cause interference problems, even though for voice services that could be still acceptable (with C/I = 9 dB).

This method does not affect other network elements or interfaces and is a very simple and quick for implementation. Also it does not affect signaling because the reselection is performed when the MS is at the idle mode.

Nevertheless, the operator has to know the exact radio propagation model to set the correct values on the cell that is congested. Finally, the operator has to predict the traffic congestion in the specific cell for best possible performance.

2.5.5.2 Expected Results

The following example shows the dynamic modification of the C2 value to resize congested cells for the elimination of the experienced congestion. Cell resizing with the use of C values is calculated, using the "WINPROP" planning simulator [41].

Let's assume that there is an urban area, covered by a network with a scheme of 3/9. This means that we have three BTSs with three sectors (cells) on each, which use the same group of frequencies. Such planning yields very good reuse of the frequencies with the use of directional antennas in each sector. The C/I level for this planning is about C/I = 17 dB.

The normal planning of the network before applying the technique is shown in Figure 2.35, where the different gray-scales show the serving area of each cell.

The highlighted area in the square box in Figure 2.35 is the congested cell. By applying the technique at the congested cell with following parameters: CellReselectionOffset = 6 dB, $T_{pen} = 640$ sec, and C2 = C1 − 6 dB, the coverage for idle users is as depicted in Figure 2.36.

The new serving area of the congested cell has minimized its surface by half compared to the previous planning. This means that if we have uniform distribution of the users, the new resized cell has half of the offered load from previous conditions.

Also if those two values are changed to: CellReselectionOffset = 12 dB and $T_{pen} =$ 640 sec, this yields C2 = C1 − 12 dB; the results are shown in Figure 2.37.

In the same way, the new serving area of the congested cell has minimized its surface by a quarter of what was covered in the normal planning or a quarter of the

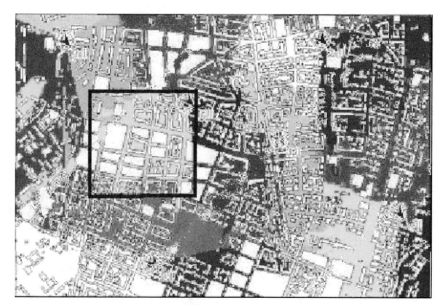

Figure 2.35   Cell-breathing effect in idle mode.

Figure 2.36   Network planning with C2 values implemented (C2 = C1 – 6 dB).

offered traffic load of the previous conditions. It should be noted that the rest of the traffic load is distributed to the adjacent cells of the congested cell.

Numerical examples.   If the blocking probability of a cell is more than 10%, the parameters are set as follows:

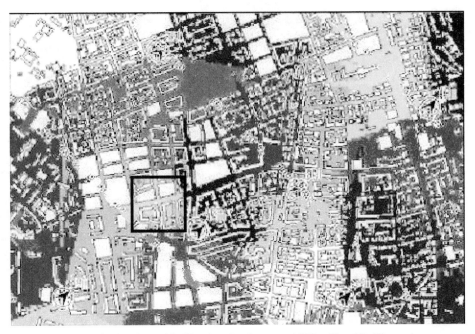

Figure 2.37 Network planning with C2 values implemented (C2 = C1 – 12 dB).

$$\text{CellReselectionOffset} = 6 \text{ dB}, \ T_{\text{pen}} = 640 \text{ sec}$$

If the blocking probability of a cell is more than 20%, the parameters are set as follows:

$$\text{CellReselectionOffset} = 12 \text{dB}, \ T_{\text{pen}} = 640 \text{ sec}$$

The introduction of this hysteresis (6 dB – 12 dB) between these values (10–20%) aims to eliminate the domino effect in the C2 value technique. The domino effect results in an extremely uncontrolled situation where cell reselection is not focused only in the radius of two or three cells but expands to the whole network.

We will further explain the previous examples, where each of the above cells has 12 TCHs and 4 SDCCHs (combined configuration). Let us assume there are 735 users camped to each of the cells, which produce 8 mErlang respectively. The total produced telecommunication traffic from the users is 5.88 Erlang. The offered traffic of 5.88 Erlang with 12 TCH configuration results in a blocking probability of 1%. In addition, for the cell in the square with channel (frequency) 20, the offered traffic load is doubled, because new users are camped to the cell or the already existent users increase their telecommunication activities. Therefore, the new traffic of 11.76 Erlang with 12 TCHs yields from Erlang B table a blocking probability 18.9%. Considering the criteria for the blocking probability, the configuration is: CellReselectionOffset = 6 dB and T_{pen} = 640 sec. As a result the congested cell covers half of the previous area. This means that only half of the users are camped in the cell; the other half is distributed to the adjacent cells. It is remarkable to notice that the users that occupied TCHs in the congested cell, before the implementation of the technique, will remain in this cell, and they will not initiate any handover even if these users are located out of the new serving area.

The results that are produced from the traffic simulator are as follows:

- *Congested cell (channel 20):* Blocking Probability from 18.9% to 1%;
- *Adjacent cells (channels 90 and 40):* Blocking probability of 1 to 6%;
- *Adjacent cells (channels 10 and 30):* Blocking probability of 1 to 1.8 %.

Because the new values of the blocking probability of the adjacent cells are lower than the activation thresholds of 10% or 20%, the technique does not affect the adjacent cells. The mobile terminals that are camped in the new border of the congested cell and its adjacent cell will face a C/I = 15 dB compared with the 17 dB planned.

Moreover, let us assume that in the cell with channel 20, the offered traffic is tripled, as new users are camped to the cell or the already existent users increase their telecommunication activities. The new load of 17.64 Erlang with 12 TCHs gives from the Erlang B table a blocking probability of 39.1%. Hence, with the above criterion (39.1% > 20% threshold), the parameters set are CellReselectionOffset = 12 dB and T_{pen} = 640 sec, with the result that the congested cell covers one-quarter of the previous area.

The new blocking probabilities for congested and adjacent cells are listed as follows:

- *Congested cell:* Blocking probability of 39.1 to 1%;
- *Adjacent cells (channels 90 and 40):* Blocking probability of 1 to 16.7%;
- *Adjacent cells (channels 10 and 30):* Blocking probability of 1 to 3.1%.

Because the new blocking probability of the adjacent cells is 16.7%, thus 6.7% above the 10% threshold, the technique is reapplied for these adjacent cells (domino effect). The parameters for the adjacent cells are set to

CellReselectionOffset = 6 dB and T_{pen} = 640 sec

and for the congested cell the new values are

CellReselectionOffset = 12 dB + 6 dB = 18 dB, T_{pen} = 640 sec

Therefore the new blocking probabilities are the following:

- *Congested cell:* Blocking probability of 39.1 to 0.1 %;
- *Adjacent cells (channels 90 and 40):* Blocking probability of 1 to 4.7%;
- *Adjacent cells (channels 10 and 30):* Blocking probability of 1 to 10%.

The domino effect even if the blocking probability of the adjacent cells is equal to the threshold value of 10% it is not reapplied and the network stops to apply any changes to the C2 parameters of the congested and adjacent cells.

The mobile terminals that are camped in the new border of the congested cell and its adjacent cell, will face a C/I = 12 dB than the 17dB of the normal planning before the first implementation of the C2 values technique.

Another criterion to uniformly distribute the traffic load between the cells, (adjacent and congested), is introduced as follows: If the offered traffic load of the congested cell has to be multiplied by α, that is: $\alpha = 2$ (doubled) or $\alpha = 3$ (tripled), for example, then the value of the CellReselectionOffset is calculated from the formula:

$$A = 20 \cdot \log\left(\frac{1}{1 - 10 \cdot \beta}\right), \beta = \frac{a - 1}{14a}$$

Thus the new GOS has been calculated as the following:

- *Congested cell*: Blocking probability of 18.9 to 3.9%;
- *Cells (channels 90 and 40)*: Blocking probability of 1 to 3.9%;
- *Cells (channels 10 and 30)*: Blocking probability of 1 to 2.1%.

It is known that the umbrella cells are used to cover gaps in network planning that are created in some areas when a BTS is lacking. The umbrella cells' range could be up to 35 km (67 km with the use of extend TRX), because they are situated at highly elevated places (mountains and hills). That is why those types of cells transmit using a large amount of power to cover such situations. These power transmissions cause problems, whereas in areas that have a local BTS, the signal that is transmitted from the umbrella is sometimes greater than the local signal, and the umbrella cell becomes dominant. Hence the umbrella cells can become congested, because they cover overestimated areas resulting in bad network utilization due to excessive call drop rate.

One solution to this problem is to decrease the transmitting power of the umbrella BTS, but then the umbrella cell could not cover enough of the planning gaps. In this way, the umbrella cells are not effective. Another technique is to increase the tilt of the antennas, but then have the same problems as in the previous situations. However, the implementation of the C2 values method has the benefit of the usage of the umbrella infrastructure only in the areas that have not local BTSs. The parameters' configuration for the C2 values technique is: CellReselectionOffset = 6 dB and T_{pen} = 640 sec, resulting to the C2 = C1 – CellReselectionOffset, which is C2 = C1 – 6 dB. In this case the local BTSs are the most preferable and dominant in their areas to support the locally produced telecommunication traffic.

In the following example, the effectiveness of the technique is discussed for another situation. Let us suppose that we have a central street that is cut vertically by LA1 and LA2, as shown in Figure 2.38.

Let us assume that the dark cell is congested. That means that the mobiles coming from LA1 to LA2 will initiate the LU procedure, but because of the congestion in SDCCH, the network will face a large number of LU failures.

The cells that are located along the road have a coverage of about 1 km, and the average speed of the vehicles in this road is 60 km/h; therefore a mobile has 20 sec to cross the congested cell. All the MSs, crossing the congested cell, must ignore this cell and initiate the LU procedure in its adjacent cells.

The implementation of the C2 values technique in that case is as follows:

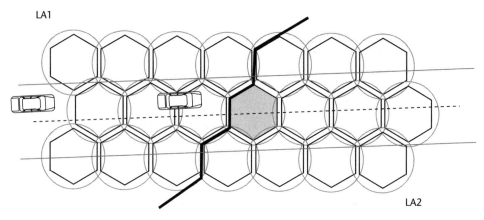

Figure 2.38 LA1 to LA2 crossing over congested cell.

$$C2 = C1 + \text{CellReselectionOffset} - \text{TemporaryOffset} \cdot H\left(T_{\text{pen}} - T\right)$$

$$\text{TemporaryOffset} = 12\,\text{dB}$$

$$T_{\text{pen}} = 20\,\text{sec}$$

$$\text{CellReselectionOffset} = -6\,\text{dB}$$

With these settings the MS that is crossing the border of the LAs via the congested cell with the speed of 60 Km/h, will ignore the congested cell, because the C2 = C1 − 6 dB − 12 dB = C1 − 18 dB, and its adjacent cells are much more attractive for becoming serving cells. The value CellReselectionOffset = −6 dB affects both the static and moving users but the parameters *TemporaryOffset* = 12 dB and T_{pen} = 60 sec affect only the moving MSs.

Also, the parameter CellReselectionOffset = −6 dB performs the resizing of the cell. On the other hand, this will cause a problem in the case of static users that belong to the congested cell and are located near the border of the LAs. With those settings these users will perform location update because of the reselection to the adjacent cells of LA1. Thus the parameter CellReselectionHysteresis of these cells has to be changed to:

CellReselectionHysteresis(old) + CellReselectionOffset = 4 dB + 6 dB = 10 dB

The penalty time will take values proportionately to the time that the congestion takes place. Thus, in heavy traffic congestion hours the penalty time will have much bigger values (e.g., 200 sec).

2.5.6 Bandwidth Reservation

This technique can be utilized to guarantee a service to a specific group of subscribers according to the traffic load situation. During the call establishment procedure, the subscriber requests a session establishment over RACH, while a random number is generated in the BSC of the system (in most vendors and suppliers). On the network side there is a table that shows the probability for call establishment, mapped to the numbers that are generated from the BSC. By dynamically changing this table it is possible to increase or decrease the probability for a call set up in a specific group. In addition, it is possible to have different tables for different groups of users. In a catastrophic event, one solution would be to enable a 100% call set up probability for the authorities, while minimizing the probability for the civilians. In addition, in extreme congestion situations it is better to drop a call at this early stage, instead of proceeding to the assignment of signaling resources [40].

2.5.6.1 Resource Management Technique Parameters

The parameters of this resource management technique mainly focus on the dynamic changing of probabilities that should be given to the group of users for the call establishment procedure. This of course should be decided based on the traffic load scenario that is predicted to apply in the specific congestion situation.

2.5.6.2 Expected Results

We assume that there is a BTS with 20 TCH, where the average offered traffic is 15 Erlang. This means that there are only five idle channels in this BTS. Thus we calculate the blocking probability to be 4.7%, using the Erlang B table, for 15-Erlang offered traffic and 20 available TCHs. However, the idle TCHs are five, and these will not be allocated with the same probability to the users of different classes, as these can be defined with this management technique.

Assume that users of the A traffic type are 98% of all the user population of the network and the rest 2% are users of type B. To perform the calculations, we define a random number, "dice," generator as D and the threshold as T. Thus, the blocking probability for the A users will now be:

$$0.98 \cdot 15 = 14.7 \text{ Erlang}$$

The probability of having $D < T$ to proceed with the allocation of a TCH is

$$P_{D<T} = \frac{70}{101} = 0.693$$

$$GOS = 1 - P_{D<T} \cdot (1 - GOS(14.7)) = 1 - \frac{70}{101} \cdot (1 - 0.041) = 1 - 0.693 \cdot 0.959 = 33.5\%$$

where, GOS(14.7) is the blocking probability for 14.7 Erlang produced from mobile users and the total number of 20 TCH in a BTS.

The blocking probability of the B users will now be:

$$0.02 \cdot 15 = 0.3 \text{ Erlang}$$

The probability of having $D < T$ to proceed with the allocation of a TCH is
$$\frac{101}{101} = 1$$

$$GOS = 1 - P_{D<T} \cdot (1 - GOS(0.3)) = 1 - \frac{101}{101} \cdot (1 - GOS(0.3)) = 1 - 1 \cdot (1 - 0.0) = 0\%$$

Let us suppose now that we have a BTS with 20 TCHs. In this BTS the average offered traffic is 19 Erlang. This means that there is only one TCH idle in this terminal. So from the Erlang B table, for 19-Erlang traffic and 20 available TCHs, the blocking probability is calculated to be 13.4%. In the same way, the only idle TCH will not be allocated with the same probability to the users of A and B class. We make again the assumption that the users of the A traffic type are 98% of all the user population of the network and that the remaining 2% are users of type B.

Therefore, the blocking probability of the A users will now be:

$$0.98 \cdot 19 = 18.62 \, \text{Erlang}$$

The probability of having TCH is $\frac{10}{101} = 0.099$

$$GOS = 1 - P_{D<T} \cdot (1 - GOS(18.62)) = 1 - \frac{10}{101} \cdot (1 - 0.124) = 1 - 0.099 \cdot 0.876 = 91.3\%$$

The blocking probability of the B users will now be:

$$0.02 \cdot 19 = 0.38 \, \text{Erlang}$$

The probability of having $D < T$ to proceed with the allocation of a TCH is $\frac{101}{101} = 1$

$$GOS = 1 - P_{D<T} \cdot (1 - GOS(0.38)) = 1 - \frac{101}{101} \cdot (1 - GOS(0.38)) = 1 - 1 \cdot (1 - 0.0) = 100\%$$

2.5.7   Location Update

Location management schemes are essentially based on users' mobility and incoming call rate characteristics. The network mobility processes face strong competition between its two basic procedures: location and paging. The location procedure allows the system to keep the knowledge of the user's location, more or less accurately to be able to find it, for example, in the case of a, incoming call [40].

Nowadays, the location method most widely implemented in 1G and 2G cellular systems (e.g., NMT, GSM, and IS-95) makes use of LAs. In these wide area radio networks, location management is done automatically. LAs allow the system to track the users during their roaming in the network. The subscribers' location is known if the system knows the LA in which the subscribers are located. When the system is about to establish communication with the mobile, paging occurs only in the current LA where the user is located. Thus, resource consumption is limited to this LA; paging messages are only transmitted in the cells of this particular LA. The location update procedure is divided into periodic and nonperiodic location updates. Periodic location updates aim to keep a back up of the system's status in case of network failure, but on the other hand it increases the congestion on the system resources and especially on SDCCH.

The optimum LA planning will save SDCCH resources from an unnecessary traffic load that will be produced by the ping-pong type of LU. Also the border of the LA could be changed only in very low traffic hours, because when we attempt a change to the LA dimensions the network will have to accept a large number of LUs. The only parameter that it is possible to change during congestion situations is the periodic location update counter.

The only effect on the network because of the location update is the use of signaling resources (SDCCH). Also the paging load is linked directly with the periodic location update. Consequently, when the time for periodic location update is high, the SDCCH load decreases, while the load on paging channels increases. Thus the operator has to define the trade-off between SDCCH load and PCH load.

2.5.7.1 Resource Management Technique Parameters

This method can be easily applied, because it only requires that the mobile periodically transmit its identity to the network. Its drawback is the associated resource consumption, which is user-dependent and can be useless if the user does not move from a LA for several hours. The location update is periodic in time and is carried out by the MS. It is used to check whether the location information in MSC/HLR is correct, because due to errors in the MSC/HLR, the location information of the MS can disappear. A specific timer function controls the periodic location update.

2.5.7.2 Expected Results

For the periodic location update it has been shown that the only way to change this effect and to minimize the congestion that is produced is from the timer, as described above. This value has been changed four times from the initial value. As a result we notice a significant decrease in location updates, while the LU success rate decreased slightly from 99.8% to 99.5%, as shown in Figure 2.39.

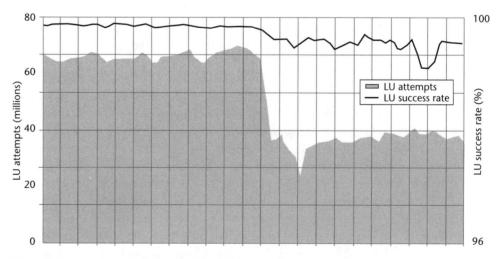

Figure 2.39 Location update attempts—location update success rate.

On the other hand, the impact on paging was a decrease of about 0.5% on the success rate, which can be interpreted as a negative effect of the change, as shown in Figure 2.40.

However, this change aims to decrease the congestion that arises in SDCCH channel, which is more important. It is known that call request and SMS request as well as location update are the main causes for seizing an SDCCH channel; therefore, the change should have a positive impact on SDCCH performance, as shown in Figure 2.41.

As a result, this leads to a decrease in the total SDCCH traffic of more than 10% as shown in Figure 2.42.

2.5.8   Time-Limited Calls

The time-limited call is an implementation of the forced release of an ongoing call, with the criterion being the duration of the session. Therefore, if the operator enables this feature, a user that has a telephone session or data session above the defined time duration, then the forced release feature initiates the release of a TCH resource. This feature should be implemented in situations of heavy communication traffic. In that way, all the users will have the opportunity to set up a call, but this is going to be time-limited. A counter keeps a record of the duration of a call and when the call exceeds the time limit set a forced release will be activated. Such counters already exist in the billing centers, while at the same time there is an exception for the emergency calls, which are excluded from the billing procedure [40].

2.5.8.1   Resource Management Technique Parameters

The only parameter of this technique is the maximum time limit for a call's duration. The operator has to set a fair value to avoid any annoyance originated by the users. An advantage of this feature is that equal opportunity is given for a call set up, since in the case that one holds a TCH for a long time, the reasoning of extreme congestion situations will force the call to be released. In the chapter on business models the effectiveness of this technique is described in detail.

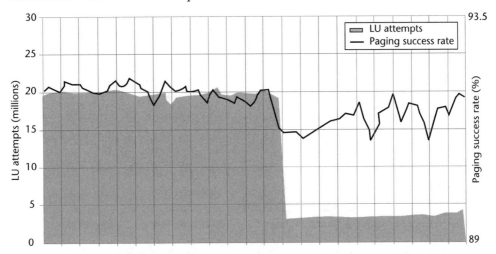

Figure 2.40   Paging success rate—location update attempts.

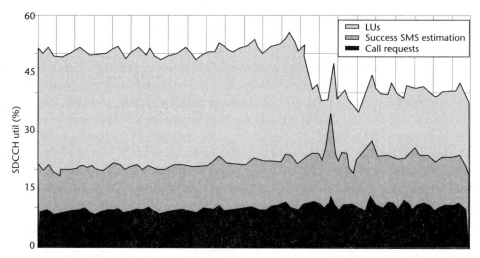

Figure 2.41   SDCCH causes share.

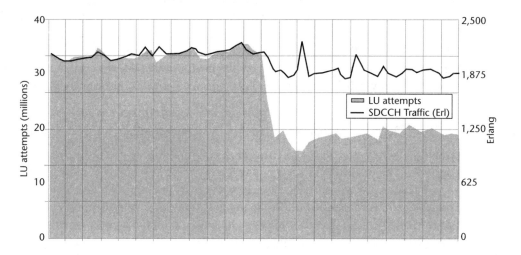

Figure 2.42   Location update attempts—SDCCH traffic.

A serious disadvantage of this feature is that it causes SDCCH congestion due to a large number of released calls, since the user whose call is released will try again to set up the call. This will cause an increased load in call set up attempts and generally in channels related to the seizure of access resources like RACH and AGCH. Also the utilization of the network will be dramatically decreased.

2.5.8.2   Expected Results

The technique aims to yield a general availability of the TCH resources in the network. By altering the subscribers' parameters this goal is achieved, as explained above, as the network maximizes the percentage of the availability of TCH channels. The trade-off between the time limitation of mobile-originating calls and the overhead signaling created from the call release is based on the number of users attached to each cell or BTS.

2.5.9 FACCH Call Set Up Due to SDCCH Congestion

When the feature of FACCH call set up is activated, it allows the MS to be assigned capacity from the TCH with the procedure of the immediate assignment, during SDCCH congestion. The TCH is then used both for signaling and traffic channel. FACCH call set up may be used in heavy SDCCH congestion situations, when band signaling is implemented instead of out-band signaling over SDCCH. If an attempt to allocate an SDCCH fails, the BSC should immediately try to allocate a TCH, provided that the FACCH call set up is allowed by an indicator in the SDCCH request message [2].

The SDCCH request does not contain any normal allocation criteria for the TCH allocation. The TCH allocation is based exclusively on the request information that is transmitted over the SDCCH. The request provides useful information, such as whether a TCH/HR is sufficient for the requested call or if a TCH/FR is required. If a TCH/HR is sufficient, an attempt is made to allocate this channel. If the TCH/HR allocation does not succeed, then an attempt is made to allocate a TCH/FR. If the request type indicates that a TCH/FR is to be allocated, then only this rate will be selected.

A TCH/FR is always allocated if an emergency call is requested, since TCH/HR cannot serve if FACCH call set up is applied. In the case of a FACCH call set up, no information of the preferred speech coding is available; therefore, the basic speech encoding algorithm of the selected channel rate is always chosen.

The A interface circuit type information is also not available in this signaling phase. Therefore, during FACCH call set up, the A interface circuit is assumed to support the TCH rate requirement, given in the SDCCH request message.

This feature can be applied also simultaneously with dynamic SDCCH. However, the FACCH call set up is only used during SDCCH congestion, when no more dynamic SDCCH channels can be assigned in the BTS. Also when the last TCH resource of the BTS is going to be occupied and the next connection requires a TCH, one is recommended to use the FACCH call set up when allowed instead of reconfiguring the TCH to the SDCCH resource.

2.5.9.1 Resource Management Technique Parameters

The operator can enable/disable the following parameters:

- Emergency calls over FACCH;
- Paging call over FACCH;
- Normal calls set-up over FACCH;
- Call set up reestablishment over FACCH.

2.5.9.2 Expected Results

We will show the expected results through an example. Let us assume that there is a BTS with 23 TCHs and eight SDCCHs. In the case of SDCCH congestion we have the following results:

- Blocking probability TCH: 9.101%;
- Blocking probability SDCCH: 28.147%;
- Blocking probability call set up: 34.687%.

In the case of implementation of the dynamic SDCCH/8, we have the following results:

- Blocking probability TCH: 11.26%;
- Blocking probability SDCCH: 0.573%;
- Blocking probability call set up: 11.768%.

In the case of implementation of the dynamic SDCCH/8 with two channels, the results are as follows:

- Blocking probability TCH: 14.4%;
- Blocking probability SDCCH: 0.003%;
- Blocking probability call set up: 13.512%.

In the case of implementation of the FACCH, the results are as follows:

- Blocking probability call set up: 9.101%;
- Blocking Probability SDCCH: 0.431%.

2.5.10 Adjusting the Rx Level

This method aims to reduce the offered traffic load to a congested cell by adjusting the appropriate values at the Rx level parameter. The RxLevelMinAccess is the minimum power level an MS has to receive from the BTS before it is allowed to access the cell. This value is implemented from the operator, and the values depend on the type and the role of the antenna. Thus, when this value is reduced, a number of MSs that do not fulfill the Rx level criterion have not had the opportunity to access the network. So this method preempts a number of users with the only criterion being their location and their distance from the cell. This method could be implemented in two ways with different results:

1. The first feature is to increase the RxLevelMinAccess only. This feature works as the C2 cell resizing methods but with one disadvantage that we will see below.
2. The second feature is to increase the RxLevelMinAccess with A dB and set the parameter CellReselectOffset = A dB. Then the method will ban the access to the MSs that do not fulfill the RxLevelMinAccess criterion.

Therefore, in the situation of heavy traffic distributed to the entire network, as in earthquakes, the operator has to implement the second feature with an appropriate value to achieve the best utilization from the network resources.

2.5.10.1   Affected Network Elements

This method has no affect to other network elements. The only effect is on network planning. In situations that we have to reduce the coverage area of a congested cell, we might have a decrease in the C/I ratio of about 5 dB in the worst case. Also for cells that are located at the border of two different LAs, different settings are selected for cell reselection hysteresis to prevent useless LUs.

2.5.10.2   Resource Management Technique Parameters

RxLevelMinAccess is the minimum receivable signal level in which the mobile terminal may use the cell. This value is implemented from the operator and the values depend on the type and the role of the antenna.

When the Rx Level is increased for A dB then the covered area in square units is reduced for $10^{\frac{A}{20}}$. Hence choosing the optimum A, with the target being the 1% blocking probability, allows the best network utilization.

When the operator enables the Rx level technique, the network topology should be planned with overlapping areas between cells. On the other hand, when the size of the congested cell becomes very small due to very big values to cell reselection offset then the MSs that are located in the new border of the congested cell will face up a C/I with 5 dB less when compared with the normal planning. Thus if the normal planning is very marginal with C/I = 11 dB, such a degradation in the C/I will cause interference problems, even though for voice services this could be acceptable with C/I = 9 dB.

2.5.10.3   Expected Results

Let us assume that there is a cell with RxLevelMinAccess = –100 dBm. Then this cell will face a blocking probability of 0.63%. If the RxLevelMinAccess is set to –102 dBm the cell will face up a blocking probability of 1%.

Consequently, the network works with the best utilizations. It should be noted that in the above example the suggestion was that there be a uniform distribution of the users in the area of the coverage of the cell. However with real time (feedback) measurement and computer-based decision it will be possible to define the optimum value.

2.5.11   Modifying BCCH Frequency List Due to Traffic Congestion

When a MS is in idle state, the cell selection and reselection is handled from the frequency list of the adjacent BCCH frequencies that are transmitted from the BCCH channel of the serving cell. The BCCH frequencies that are candidates for reselection are sorted based on the comparison of C2 values. Thus, the cell with the bigger

C2 value after the C2 value of the serving cell is the first candidate for reselection. Also this list is valid not only in idle mode but when a call is taking place. The BCCH frequency list could contain up to 32 frequencies. This list is transmitted from the MS via SACCH, during a session with the BTS and defines which cell is the first candidate for handover [40].

Particularly, the list is sent to the MS in system information message type 2 on the BCCH, and the MS may store it when powering down. Now the MS does not need to search through the whole band when powering up. If the list contains all the BCCH carriers of a certain geographical area of a PLMN, the MS can use it to search for the suitable radio frequency channels quickly to camp on a cell.

When a cell is congested the first thing that the operator has to do is to decrease somehow the offered telecommunication traffic because in this way the utilization of the network will hold at an acceptable level. One way to achieve this is to reduce the number of the incoming MSs, which are going to camp in the congested cell. Thus, a congested cell has to be buffered, and a new MS should not be served from this cell. An efficient way to do this is that the operator can modify dynamically the BCCH frequency lists that are transmitted from neighboring cells. Thus the congested cell is going to be excluded from the new list, and so the problematic cell will no longer be a candidate for reselection or handover. As a result the offered telecommunication traffic for the congested cell will be decreased, and it will be distributed to the nearby cells. This method acts like the dynamic cell resizing but only for the users that are under movement. That means that the already camped MSs into the congested cell will remain in this cell.

The handover failure rate may be increased if the network planning has inadequate overlapping. When operators use the combined channel mapping with one SDCCH/4 on the broadcast radio time slot, then to enable the broadcast messages they have to sacrifice these signaling resources for the broadcast transmission.

2.5.11.1 Resource Management Technique Parameters

The only parameter for this technique is the frequency list of the BCCH, which has to be modified. The operator has to arrange the number of sequential located removed cells, due to congestion, because this will cause reduction to the C/I and inadequate radio covered areas. So the compensation will be good for the number of adjacent cells that are going to be removed. This implementation is valid for small-scale congestion situations.

This technique has the advantage that the operator can dynamically cut off a congested cell without producing handovers or call set up failures. Also the only care during the network planning is that there should be adequate overlapping between the cells of the network.

A serious disadvantage is that if the network planning does not care for adequate overlapping, then the implementation of this method will produce black spots in the deployed network.

Also its efficiency is strongly affected by the mobility of the users. If there is a large number of static users, then the dynamic cell resizing has to be implemented. Also the already camped MSs will remain in the congested cell. That means that the method is not efficient for users with small mobility.

2.5.11.2   Expected Results

We will examine what happens by applying this method through a suitable example as shown in Figure 2.43. Suppose that the black line shows a central street that discriminates LA1 from LA2. The condition is that the dark cell has been congested (e.g., SDCCH congestion). That means that the MSs, which come from LA1 to LA2, will initiate the LU procedure, but because of the congestion, the network will face a big number of LU failures.

The operator has to isolate the congested cell from the other network and reduce the newly camped MSs. Thus the operator modifies the BCCH frequency lists of the adjacent cells of the congested one, in a way that will not include the BCCH frequency of the congested. Thus, a MS that is passing from LA1 to LA2 as in Figure 2.43 is going to ignore the congested cell and the LU is going to happen at the adjacent cells of the congested one. Figure 2.44 shows an example, where the cell with the dark color is congested and the adjacent cells transmit the BCCH frequency after applying the technique, so that a mobile subscriber crossing the area will not "consider" this cell.

2.5.12   Other Resource Management Techniques

This section briefly presents other resource management techniques that can be applied to cellular systems of the 2G.

2.5.12.1   Enhanced Multilevel Precedence and Preemption (eMLPP)

The eMLPP standard defines different priority levels to call requests, based on parameters for call set up and continuation of a connection during the handover procedure. Using this priority feature, the network can release or queue certain connections in the air interface during congestion situations provided that this has been defined at the beginning of the call [42, 43].

According to the GSM standards, priority can be applied in two procedures for traffic channel allocation, namely the assignment and handover. The eMLPP standard, defines two parts:

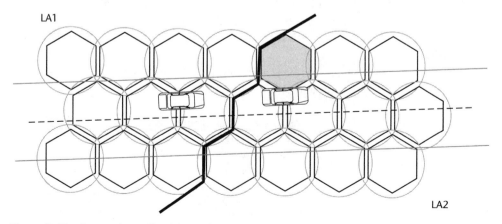

Figure 2.43   Congestion cell at LA crossing.

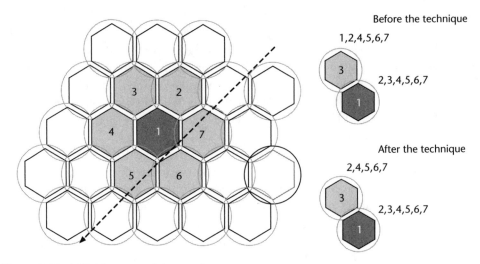

Figure 2.44 BCCH frequency list example.

1. Precedence, by assigning a higher priority level to a call request;
2. Preemption, by reserving resources for a higher level precedence call in the absence of idle resources.

The eMLPP should be also applicable in the case of roaming, if this is supported by the serving networks. The maximum precedence level of a subscriber is defined during the subscription by the service provider, based on the subscriber's contract. The subscriber should be able to select a precedence level up to the maximum level subscribed to, on a per call basis (provided that he or she has an eMLPP-compatible mobile terminal).

The seven eMLPP priority levels.   The priority levels A and B are the highest, and they are reserved for network internal use (e.g., emergency calls, network-related services, voice broadcasts, or voice group call services). These two priority levels should only be used locally (i.e., in the domain of one MSC). The other five priority levels are offered for subscription and can be applied globally (e.g., on interswitch trunks) if this is supported by all related network elements. Moreover, they can be applied for networking with ISDN networks providing the MLPP service. The seven priority levels are defined in Table 2.9.

Levels A and B should be mapped to level 0 for priority treatment outside of the MSC area in which they are applied.

Table 2.9   eMLPP Priority Levels

Level A (highest/network internal use)
Level B (network internal use)
Level 0 (subscription)
Level 1 (subscription)
Level 2 (subscription)
Level 3 (subscription)
Level 4 (lowest/subscription)

2.5.12.2   Queuing

The purpose of radio resource queuing in the BSC is to increase the number of successfully completed calls in a temporary congestion situation in the BTS and thereby increase radio network efficiency. Radio resource queuing enables the setting of the radio channel request to the queue, and when a suitable radio resource is available again, the interrupted call set up can be continued. Consequently, there is no need to cut off started transactions facing temporary radio channel congestion in the actual BTS. The queued seizure request can be a call set up, MOC set up, or a mobile terminating call (MTC) set up, or a handover attempt (all GSM-specified handover types). The queuing technique contains specific priority management and statistical functions [40].

Queuing is used when the BSC receives an assignment or a handover request from the MSC and all traffic channels are busy, or there are no TCHs of the requested kind available. If this seizure request is not allowed to preempt an existing connection, the BSC checks the priority information element and, on the basis of the "queuing allowed" indicator, it may initiate queuing.

2.5.12.3   Directed Retry

Directed retry is another procedure that is can be used when congestion is detected in the network, where the mobile terminal is assigned to a traffic channel in a cell other than the serving cell during the call set up phase. This feature is applied when the network does not have any free TCHs for allocation. In that case the relevant BSC that is informed from the MS about which cells receive the MS, searches for free TCHs in these adjacent cells to proceed with the TCH allocation [40].

By means of a directed retry procedure it is possible to avoid the drop of a call if the serving cell is temporarily congested. The directed retry is triggered by an assignment procedure and can be used either in a mobile-originating call or a mobile-terminating call attempt. The directed retry function involves MSs, BSCs, and MSCs.

2.5.12.4   Access Control by User Groups

In extreme congestion situations when most of the resource management techniques do not solve the problem, the access control by user groups can be utilized to provide a practical solution for the congested network. Whenever resource allocation cannot be applied for reasons like the scale of the situation, the solution offered by this technique is to decrease the number of the mobile terminals that "compete" to set up calls. The network activates the feature of access control by user groups to map the available network resources to a call arrival rate that can be served by the network [40].

The system broadcasts through the BCCH the list of authorized access classes and authorized special access classes, and whether emergency calls are allowed in the cell to all mobile stations or only to the members of authorized special access classes.

The access control technique minimizes the overhead of congested cells by selecting a group of subscribers randomly and denying their access. This results in a decrease in the subscribers attached in each BTS and at the same time maximizes the percentage of all the network resources to the specific subscribers' classes that gain access to the network.

2.5.13 Evaluation Criteria for the Resource Management Techniques Selection

The resource management techniques that have been described in the previous sections, aim to tackle congestion in GSM networks. This section focuses on the joint evaluation and comparison of these techniques according to the scale, the logical channels, the extra congestion on adjacent cells, and the user satisfaction related to each technique.

2.5.13.1 Resource Management Technique Versus Scale

Each technique is described with an efficiency factor according to the scale of the congestion situation. The scale of the phenomenon is then divided into time (duration of a congestion situation) and place (number of congested cells). Thus each technique results into different situation, and Table 2.10 summarizes the techniques.

2.5.13.2 Resource Management Techniques and Blocking in Logical Channels

In all aspects of the resource management techniques presented the aim is to decongest different network resources. Generally, to solve the congestion problem three alternatives exist: The first one is to add extra resources from network point of view (adding new TRX in the network); the second one is to reallocate existing resources by converting TCH channels to SDCCH or changing the channel mapping like in the BCCH converting the total number of AGCHs and PCHs. The third technique is to cut off the resources from a number of users, which means that some users will be banned from accessing the network in order to help other subscribers. It is obvious that the most efficient way is to reallocate these logical channels—resources

Table 2.10   Management Techniques Versus Congestion Scale

	Scale					
	Time (Duration of Congestion Situation)			Place (Number of Congested Cells)		
Techniques	*Low*	*Medium*	*High*	*Low*	*Medium*	*High*
Half rate	x	x	x	x	x	x
Forced handover	x	—	—	x	—	—
Dynamic SDCCH	x	x	x	x	x	x
TRX prioritization	x	x	x	x	x	x
Dynamic cell resizing	x	x	x	x	x	—
Bandwidth reservation	x	x	x	x	x	x
eMLPP	—	—	—	—	—	—
Location update	—	x	x	x	x	x
Queuing	x	—	—	x	x	—
Time-limited calls	—	—	x	—	—	x
FACCH	x	x	x	x	x	x
Adjusting the Rx level	—	—	x	—	—	x
Directed retry	x	—	—	x	—	—
User groups	—	x	x	—	x	x
BCCH frequency list	x	x	x	x	—	—

dynamically to the network and the subscribers since adding more resources costs and banning some users leads to complaints and a bad reputation for the operator. These logical channels are: SDCCH, AGCH, PCH and TCH channels. The results are shown in Table 2.11.

2.5.13.3   Resource Management Techniques and Implications to the Adjacent Cells

When a technique is applied, the resources that are allocated are sometimes borrowed from the resources of the adjacent cells. According to the blocking probability for the adjacent cells these resources should be handled in an appropriate manner. In addition, and respective to the applied technique, the adjacent cells may become congested. This possibility is shown in Table 2.12.

Table 2.11   Effectiveness in Exploiting Logical Channels During Congestion

Techniques	Logical Channel (Blocking)			
	AGCH	PCH	SDCCH	TCH
Half rate	—	—	—	x
Forced handover	—	—	x	x
Dynamic SDCCH	—	—	x	—
TRX prioritization	—	—	—	x
Dynamic cell resizing	x	x	x	x
Bandwidth reservation	x	—	x	x
eMLPP	—	—	x	x
Location update	—	x	x	—
Queuing	—	—	—	x
Time-limited calls	—	—	—	x
User groups	x	x	x	—
BCCH frequency list	x	x	x	x
FACCH	—	—	x	—
Rx level	x	x	x	x
Directed retry	—	—	—	x

Table 2.12   Adjacent Cell Implications

Techniques	Adjacent Cells (Becoming Congested)
Half rate	—
Forced handover	x
Dynamic SDCCH	—
TRX prioritization	—
Dynamic cell resizing	x
Bandwidth reservation	—
eMLPP	—
Location update	—
Queuing	—
Time-limited calls	—
User groups	—
BCCH frequency list	x
FACCH	—
Rx level	x
Directed retry	x

Table 2.13 User Satisfaction

Techniques	*User Satisfaction*
Half rate	x
Forced handover	x
Dynamic SDCCH	x
TRX prioritization	x
Dynamic cell resizing	x
Bandwidth reservation	—
eMLPP	—
Location update	x
Queuing	—
Time-limited calls	—
User groups	—
BCCH frequency list	x
FACCH	x
Rx level	—
Directed retry	x

2.5.13.4 Resource Management Techniques and User Satisfaction

From the network point of view, all the techniques result in better resource utilization. Even if the outcome, from the implementation of a technique is quite marginal, it will still maximize the efficiency of the available resources. On the other hand, users without the technical knowledge in network planning and network operation always characterize a network by their satisfaction with using the network services. Table 2.13 shows the effects on user satisfaction versus the techniques after their implementation.

References

[1] Mouly, M., and M. Pautet, "The GSM System," France, 1992.

[2] Walke, B. H., "Mobile Radio Networks: Networking and Protocols," New York: John Wiley & Sons, 1999.

[3] http://www.3gpp.org.

[4] http://www.3gpp.org/specs/status.htm.

[5] http://www.3gpp.org/specs/releases.htm.

[6] 3GPP, TSG GSM/EDGE, "Radio Access Network: Multiplexing and Multiple Access on the Radio Path," 3GPP TS 45.002 (V4.6.0), February 2003.

[7] 3GPP, TSG GSM/EDGE, "Radio Access Network: Digital Cellular Telecommunications System (Phase 2+); Modulation," 3GPP TS 45.004 (V4.2.0), November 2001.

[8] 3GPP, TSG GSM/EDGE, "Radio Access Network: Radio Transmission and Reception," 3GPP TS 45.005 (V4.10.0), February 2003.

[9] 3GPP, TSG GSM/EDGE, "Radio Access Network: Radio Subsystem Link Control," 3GPP TS 45.008 (V4.9.0), June 2002.

[10] 3GPP, TSG GSM/EDGE, "Radio Access Network: Link Adaptation," 3GPP TS 45.009 (V4.2.0), November 2001.

[11] 3GPP, TSG GSM/EDGE, "Radio Access Network: Radio Subsystem Synchronization," 3GPP TS 45.010 (V4.4.0), February 2003.

[12] ITU-T Recommendation X.200, "Information technology—Open Systems Interconnection—Basic Reference Model: The Basic Model," July 1994.

[13]   3GPP, TSG GSM/EDGE, "Radio Access Network: Layer 1; General Requirements," 3GPP
        TS 44.004 (V4.2.0), December 2001.

[14]   3GPP, TSG GSM/EDGE, "Radio Access Network: Data Link (DL) Layer: General
        Aspects," 3GPP TS 44.005 (V4.0.0), April 2001.

[15]   3GPP, TSG GSM/EDGE, "Radio Access Network: Mobile Station—Base Station System
        (MS-BSS) Interface: Data Link (DL) Layer Specification," 3GPP TS 44.006 (V4.1.0), Febru-
        ary 2002.

[16]   3GPP, TSG Core Network, "Mobile Radio Interface Signaling Layer 3: General Aspects,"
        3GPP TS 24.007 (V4.2.0), June 2002.

[17]   3GPP, TSG Core Network, "Mobile Radio Interface Layer 3 Specification: Core Network
        Protocols: Stage 3," 3GPP TS 24.008 (V4.10.0), March 2003.

[18]   3GPP, TSG GSM/EDGE, "Radio Access Network: Base Station Controller—Base Trans-
        ceiver Station (BSC-BTS) Interface: General Aspects," 3GPP TS 48.051 (V4.1.0), December
        2001.

[19]   3GPP, TSG GSM/EDGE, "Radio Access Network; Base Station Controller—Base Trans-
        ceiver Station (BSC-BTS) Interface: Interface Principles," 3GPP TS 48.052 (V4.0.1), May
        2001.

[20]   3GPP, TSG GSM/EDGE, "Radio Access Network: Base Station System—Mobile-Services
        Switching Center (BSS-MSC) Interface: General Aspects," 3GPP TS 48.001 (V4.0.0), April
        2001.

[21]   3GPP, TSG GSM/EDGE, "Radio Access Network: Base Station System—Mobile-Services
        Switching Centre (BSS-MSC) Interface: Interface Principles," 3GPP TS 48.002 (V4.2.0),
        December 2001.

[22]   ITU-T Recommendation X.25, "Interface Between Data Terminal Equipment (DTE) and
        Data Circuit-Terminating Equipment (DCE) for Terminals Operating in the Packet Mode
        and Connected to Public Data Networks by Dedicated Circuit," October 1996.

[23]   CAUTION EU Project, IST-2000-25352, Deliverable D-2.1, "Requirement Analysis and
        Functional Specifications," July 2001.

[24]   Kyriazakos, S., et al., "Performance Evaluation of GSM and GPRS Systems Based on Meas-
        urement Campaigns and Statistical Analysis," *IEEE Globecom 2002,* Taipei, Taiwan,
        November 2002.

[25]   Kyriazakos, S., et al., "A Comprehensive Study and Performance Evaluation of Operational
        GSM and GPRS Systems Under Varying Traffic Conditions," *IST Mobile Summit 2002,*
        Thessaloniki, Greece, June 2002.

[26]   Temple, J., S. McGrath, and C. J. Burkley, "The Pan-European GSM Signaling Require-
        ments for the A_{bis} Interface," *Proceedings of the 44th IEEE Vehicular Technology Confer-
        ence,* Stockholm, June 1994, pp. 775–779.

[27]   Verdone, R., and A. Zanella, "Performance of Directed Retry in a Mobile Radio System:
        The Impact of Propagation," *ITC2000,* Acapulco, Mexico, May 2000, pp. 167–171.

[28]   Yum, T.-S. P., and K. L. Yeung, "Blocking and Handoff Performance Analysis of Directed
        Retry in Cellular Mobile Systems," *IEEE Transactions on Vehicular Technology,* Vol. 44,
        No. 3, August 1995, pp. 645–65.

[29]   ITU-T Recommendation E.750, "Introduction to the E.750 series of Recommendations on
        Traffic Engineering Aspects of Networks Supporting Personal Communications Services,"
        March 2000.

[30]   Grillo, B., "Personal Communications and Traffic Engineering in ITU-T: The Developing,
        E.750-Series of Recommendations," *IEEE Personal Communications Magazine,* Special
        Issue on Traffic and Mobility, 1996.

[31]   Bettstetter C., "Mobility Modeling in Wireless Networks: Categorization, Smooth Move-
        ment, and Border Effects," *ACM SIGMOBILE Mobile Computing and Communications
        Review,* Vol. 5, Issue 3, July 2001.

[32] Lam, D., D. C. Cox, and J. Widom, "Teletraffic Modeling for Personal Communications Services," *IEEE Communications Magazine*, February 1997.

[33] ITU-T Recommendation E.751, "Reference Connections for Traffic Engineering for Land Mobile Networks," February 1996.

[34] ITU-T Recommendation E.760, "Terminal Mobility Traffic Modeling," March 2000.

[35] Brass, V., and W. Fuhrmann, "Traffic Engineering Experience from Operating Cellular Networks," *IEEE Communications Magazine*, August 1997.

[36] CAUTION EU Project, IST-2000-25352, Deliverable D-3.3, "Resource Management Application High Level Design—CAUTION Resource Management," November 2001.

[37] Kyriazakos, S., et al.,"Congestion Study and Resource Management in Cellular Networks of the Present and Future Generations," *IST Mobile Summit 2001*, Barcelona, Spain, September 2001.

[38] Barbera, M., et al., "An Application of Case-Based Reasoning to the Adaptive Management of Wireless Networks," *Lecture Notes in Artificial Intelligence*, Springer Verlag, *6th European Conference on Case Based Reasoning*, Aberdeen, Scotland, September 4–7, 2002.

[39] Tripathi, N. D., J. H. Reed, and H. VanLandingham, "Handoff in Cellular Systems," *IEEE Personal Communications Magazine*, December 1998.

[40] CAUTION EU Project, IST-2000-25352, Deliverable D-3.1, "Resource Management Application Scenarios—Traffic Load Scenarios and Decision-Making," October 2001.

[41] WinProp Software Suite, AWE Communications GmbH.

[42] 3GPP, TSG Core Network, "Enhanced Multilevel Precedence and Preemption Service (eMLPP): Stage 2," 3GPP TS 23.067 (V4.1.1), June 2002.

[43] 3GPP, TSG Core Network, "Enhanced Multilevel Precedence and Preemption Service (eMLPP): Stage 3," 3GPP TS 24.067 (V4.1.0), June 2001.

CHAPTER 3

Resource Management in GSM Phase 2+ Wireless Systems

This chapter deals with radio resource management in GSM phase 2+ wireless systems. The chapter is structured around three sections. Section 3.1 describes the architecture and the technical characteristics of GSM 2+ systems. Apart from the GPRS, which is the most important feature of GSM phase 2+, HSCSD and EDGE technologies are described. Next, network dimensioning and performance analysis is discussed. Finally, a number of resource management techniques are described that allow the efficient handling of user data and the provision of QoS.

3.1 Architecture of GSM Phase 2+ Systems

This section describes the features of GSM phase 2+ systems, as these are standardized by ETSI and 3GPP. More specifically, an introduction to GPRS is given, as well as the description of its architecture and a presentation of the channels that carry out the packet-switched communications. Finally, HSCSD and EDGE technologies are introduced, thus covering the whole spectrum of features corresponding to GSM phase 2+ systems.

3.1.1 Introduction to GPRS

As described in Chapter 2, GSM was initially a circuit-switched system offering satisfactory voice-based communication and limited data services over a radio channels network. Initial GSM networks had a number of limitations as far as data transmission is concerned, which the GPRS tries to overcome [1]. Among these limitations are that the uplink and downlink channels are allocated during the entire call period of a session even if no data is transmitted and that a user's charges are based on the time they hold a call session. In addition, they had a limited bit rate up to 9.6 Kbps, a long lasting session set up, and very limited data services support (SMS and FAX). As technology evolves and new services are spread out mainly in the Internet community, mobile users are also becoming more demanding for the services with which they are provided. GPRS is an effort to enhance the existing mobile communications technology with the data services that users demand to address the initial GSM inefficiencies. GPRS transfers data over a mobile telephony network with a maximum theoretical speed up to 170 Kbps. In GPRS, users are assumed to be "always on line" and ready to use any service offered with no need of a dial-up modem connection.

The innovation of GPRS is the application of a packet-switched communication approach to transfer data packets between a mobile GSM network and other fixed, packet-switched–based data networks such us the Internet. Benefits gained with GPRS are better radio channel allocation since these are used only when data is sent, fairer billing based on the amount of data transferred and Web applications over the mobile network such as FTP, Web browsing, chat, e-mail, and telnet. GPRS is standardized by the ETSI [2] under the 3GPP [3] collaboration agreement.

3.1.2   GPRS System Architecture

GPRS enhances the GSM system architecture (Figure 3.1) by introducing GPRS support nodes (GSNs) and a number of new communication interfaces [4]. GSN is a new kind of network node responsible for the management of packets in GPRS. GSN nodes are interconnected though an IP-based backbone not only inside a public land mobile network (PLMN) but also over all existing PLMNs that support GPRS. In the first case GSN backbone networks provide GPRS services in the PLMN they belong to, while in the second case GSNs provide GPRS services between different PLMNs after a roaming agreement. Between different PLMNs there exist the so-called border gateways, which mostly performing security and roaming functions.

There are two kinds of GSNs; the serving GPRS support node (SGSN) and the gateway GPRS support node (GGSN). SGSN delivers data packets to and from the mobile stations in its service area. Functions such as packet routing and transfer, mobility management (attach/detach and location management), logical link management, authentication and charging are performed by SGSN. GGSN is the GPRS backbone gateway to packet data networks (PDNs). In one direction GGSN converts the packets that have arrived from the mobile stations, through the SGSN, to the appropriate format to be sent to an external PDN (e.g., the IP or X.25 packet format). In the opposite direction, GGSN accepts data packets from PDN and converts their packet data protocol (PDP) address to the appropriate GPRS address of the destination mobile terminal. It then propagates the packet to the appropriate SGSN

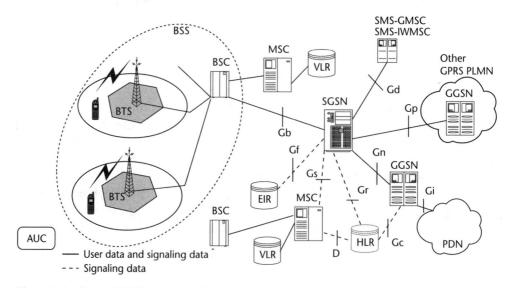

Figure 3.1   General GPRS system architecture.

where the mobile station can be contacted. To do so, GGSN stores the SGSN of each user at any time. Apart from packet conversion, other tasks GGSN performs are authentication and charging. Each GPRS-supported GSM network may have more than one SGSN covering all geographical areas the GSM network spans. It also may have more than one GGSN, each one considered as the gateway to different PDNs. Of course extensions of the functions of existing GSM components are necessary (e.g., now HLR also stores the SGSN address of a user based on his or her current location and the corresponding PDP address of each GPRS user in the PLMN).

A number of interfaces are defined for the GPRS system architecture as shown in Figure 3.1 [4]:

- The G_b interface connects BSCs to SGSN.
- The G_n interface is used between GSNs of same PLMN to exchange the user profile when the user moves from one SGSN to another.
- The G_p interface is defined between two GSNs of different PLMNs for exchanging the user profile and other signaling information between a SGSN and GGSN of another area.
- The G_f interface is used between SGSN and EIR to query the IMEI information if a MS tries to register with the network.
- The G_r interface between SGSN and HLR is used to get the user profile, the current SGSN address, and the PDP address(es) for each user in PLMN.
- The G_c interface between GGSN and HLR is used by GGSN to query the user's location and profile to update its location register.
- The G_i interface connects GGSN to external PDN (e.g., X.25 or IP).
- The G_s interface between SGSN and MSC/VLR is used to perform the paging request of a circuit-switched GSM call for combined attachment procedure.
- The G_d interface between SMS-gateway (SMS-GMSC), and SGSN is used to exchange SMS messages.

3.1.3   Air Interface of GPRS

GPRS's main advantage in relation to standard GSM is better utilization of the radio resources (radio channels). It is known that GSM allocates one slot per TDMA frame to each mobile station for a given session. On the contrary in GPRS more than one (up to eight) time slots of the same TDMA frame can be allocated to a mobile station. In GSM the same slot in sequence inside the time frame is given for uplink and downlink traffic, while in GPRS the allocation of time slots for uplink and downlink is asymmetric. In GPRS, channels are given only when data have to be transferred and not during the entire session as in GSM. Thus multiple mobile stations may use common physical channels. GPRS uses the same physical channels as the GSM inside a cell, and these are the radio channels with which a cell has been associated. These are assigned to both GPRS and non-GPRS services. The allocation of channels to services is based on the priority of a service and the current traffic load [5].

A number of logical channels are defined over the packet data channel (PDCH) of the GPRS used for signaling, broadcasting of system information, paging, and frame synchronization [6]. The packet data traffic channel (PDTCH), packet broadcast control channel (PBCCH), packet common control channel (PCCCH), and packet dedicated control channel are the basic logical channels of GPRS. Table 3.1 lists the subchannels of those basic channels with a short description about their specific use.

3.1.4 HSCSD

HSCSD is a GSM-based technology that aims to provide better bit rate transfers for the mobile stations [7, 8]. Like GSM, HSCSD is a circuit-switched technology as its name denotes. It gives the user one or more dedicated circuits for the entire call session and supports data-oriented applications that were not available to the standard GSM user of the past, such as accessing private LANs, sending and receiving e-mails, and accessing the Internet while on the move. What makes HSCSD attractive is the dedicated bandwidth it offers for time-critical data services, something that is not offered even by the newer GPRS technology. HSCSD is currently available to many operators in countries around the world, and the implementation of international roaming agreements between all HSCSD providers makes the communication of the respective mobile subscribers easier.

HSCSD was approved by ETSI in 1997 and employs two major technologies. It uses a proven channel-coding scheme that increases the channel bit rate from the existing 9.6 Kbps to 14.4 Kbps and allows the combination of channels to enable data rates in multiples of 9.6 Kbps or 14.4 Kbps [9]. This results in users being able to obtain mobile data rates ranging from 28.8 Kbps to 43.2 Kbps.

The architecture of HSCSD service is similar to that of GSM. The difference here is the options for the mobile station. While in the GSM case that was a single cell phone with just voice capabilities, in HSCSD, subscribers use either voice terminals that support the HSCSD, a portable computer that can be connected with a mobile phone (see Figure 3.2), or a special PCMCIA portable computer card, with a built in GSM phone that turns notebook computers and other portable devices into a

Table 3.1 GPRS Logical Channels

Logical Channel	Logical Subchannels	Description
PDTCH		Transfer of user data
PBCCH		Signaling channel; BSS broadcasts information about the GPRS network to mobile stations
PCCCH		Allocation of radio channels and paging
	Packet random access channel (PRACH)	Request for PDTCH
	Packet access grant channel (PAGCH)	Allocation of PDTCH
	Packet paging channel (PPCH)	Paging functions
	Packet notification channel (PNCH)	Informs mobile stations for point-to-multipoint messages
Dedicated control cannel		Signaling
	Packet-associated control channel (PACCH)	Signaling information of a mobile station that allocates resources
	Packet timing advance control channel (PTCCH)	Frame synchronization

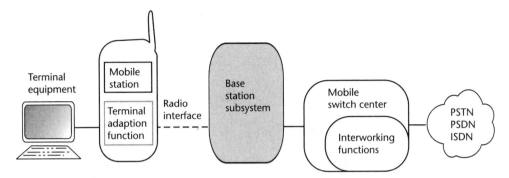

Figure 3.2   HSCSD architecture.

complete high-speed mobile office with the ability to make voice calls hands-free, as well as data transfers.

3.1.5   EDGE Technology

EDGE technology is the next step after GPRS in the enhancement of GSM [10]. The 3GPP is also responsible for the standardization of the EDGE technology. This new technology increases data transmission rates and spectrum efficiency and allows for new applications and increased capacity for mobile users. Previous services, such as HSCSD and GPRS are not modified, but enhanced by one more physical layer provided by the EDGE technology.

GPRS allows data rates of 115 Kbps and, theoretically, of up to 170 Kbps on the physical layer. Enhanced GPRS (EGPRS) offers data rates of 384 Kbps and, theoretically, of up to 473.6 Kbps. A new modulation technique and error-tolerant transmission methods, combined with improved link adaptation mechanisms, make these EGPRS rates possible. This is the key to increased spectrum efficiency and enhanced applications, such as wireless Internet access, multimedia e-mail, and file transfers.

While GPRS introduces many new protocols and new nodes over the GSM technology, EDGE just uses and enhances the existing GPRS infrastructure. EDGE can be seen as a method to increase the data rates on the GSM's radio link. EDGE introduces a new modulation technique and new channel coding that can be used to transmit both packet-switched and circuit-switched voice and data services. EDGE is therefore an add-on to GPRS and cannot work alone (Figure 3.3). By adding the new modulation and coding to GPRS and by making adjustments to the radio link protocols, EGPRS offers significantly higher throughput and capacity. GPRS and EGPRS have different protocols and different behavior on the base station system side. However, on the core network side, GPRS and EGPRS share the same packet-handling protocols and, therefore, behave in the same way. Reuse of the existing GPRS core infrastructure (serving GRPS support node/gateway GPRS support node) emphasizes the fact that EGPRS is only an "add-on" to the base station system and is therefore much easier to introduce than GPRS itself. In addition to enhancing the throughput for each data user, EDGE also increases capacity. With

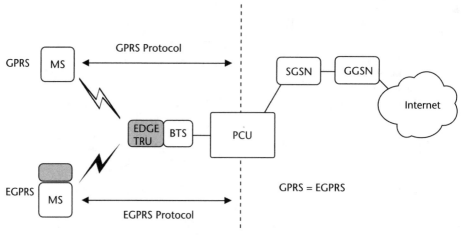

Figure 3.3 EGPRS enhancement to GPRS.

EDGE, the same time slot can support more users. This decreases the number of radio resources required to support the same traffic, thus freeing up capacity for more data or voice services. EDGE makes it easier for circuit-switched and packet-switched traffic to coexist while making more efficient use of the same radio resources.

3.2 Network Dimensioning and Performance Analysis of 2+G Wireless Systems

Network dimensioning in 2+G systems is a very important and complicated task, since this is normally carried out after the network planning and dimensioning of the circuit switched part of the system. In the case of GPRS, which was introduced many years after the original GSM system, packet-switched network planning and dimensioning is performed as a second step since it is usually carried out on an already deployed GSM network.

In GPRS systems, the QoS given to the user is most of the time linked with the average data throughput and the delay experienced. Network planning and dimensioning should provide for sufficient radio resources and QoS in order to guarantee coverage and acceptable data throughput. The GPRS planning consists of four primary steps:

- Radio coverage planning;
- System frequency planning;
- Network capacity planning;
- System parameter planning.

3.2.1 Radio Coverage Planning

For the purposes of radio coverage planning, the C/I ratio should be sufficient for the entire coverage area to guarantee the ability for successful data transmission, both on the uplink and the downlink. Each coding scheme (CS) defined for GPRS is appropriate for a particular range of C/I for a given block error rate (BLER)—that is less than 10%—which is what most of the networks require. The higher the level of error protection, the lower the required C/I for the area of coverage.

In Table 3.2, the required C/I is given under the assumption that frequency hopping is activated in the GSM network. In case this feature is not present, the required C/I can be decreased for a few decibels for most of the CSs.

The main objective for enhanced radio coverage planning is to balance uplink and downlink as far as coverage is concerned. This can be achieved by determining the link budget. Therefore, an estimation of the maximum cell range of GPRS coverage can be calculated according to the different coding schemes available. The relative coverage area of each coding scheme is different due to the different C/I. If all four coding schemes are considered, the coverage area decreases progressively as we change from CS-1, to CS-2, to CS-3, to CS-4, due to the corresponding decrease in the C/I ratio. CS-1 and CS-2 are usually appropriate for usage in the area that is covered from the circuit-switched voice network but the areas covered by CS-3 and CS-4 are reduced, since the error protection is reduced, and the C/I ratio required is higher.

Therefore, the C/I distribution in the area is the input to determine the coverage for different CSs. In some cases, it may be necessary to install additional sites to ensure GPRS coverage, since cell interference is the dominant factor for decreased GPRS performance.

3.2.2   System Frequency Planning

One of the most critical issues that cellular network planners have to face is radio interference, which is the main reason that frequency planning is required on top of the coverage planning that was presented in the previous section. Frequency reuse, a common procedure for cellular systems, influences the C/I ratio and subsequently the GPRS performance. Obviously, the denser the cells are planned, the higher the frequency reuse and the lower the resulting C/I.

Removing capacity from the already planned circuit-switched network to loosen the frequency reuse is not recommended for cellular operators, since there is a trade-off between the capacity and C/I for that area. One way to keep the system capacity constant, while the frequency reuse is minimized is to introduce micro- and macrolayers. In that case, the macrolayer can be used mainly for data communication, while the microlayer cells are offered to voice users.

The microlayer and macrolayer approach seems to be a strong candidate for operators seeking to achieve a high data communication performance. Many operators have licensed spectrum in two different bands (e.g., GSM 900 and GSM 1,800). If this is the case, even better network planning is achieved.

Table 3.2   C/I for Different CSs

CS	C/I [dB]
CS-1	~6
CS-2	~9
CS-3	~12
CS-4	~17

3.2.3 Network Capacity Planning

3.2.3.1 Calculating the Cell Configuration

In GSM/GPRS networks, voice traffic is always prioritized over packet-switched traffic, since user satisfaction is strongly linked with the ability to set up calls, rather than making use of data communication. To calculate the required number of TCHs for a given required traffic load and blocking probability, the Erlang-B formula is used:

$$B(N) = \frac{\dfrac{A^N}{N!}}{\displaystyle\sum_{i=0}^{N} \dfrac{A^i}{i!}}$$

where B is the blocking probability (percentage), A is the offered traffic load (Erl), and N is the number of traffic channels.

A capacity planning based on the above formula does not include the C/I parameter, and therefore, it is assumed that this is sufficient for a reliable transmission.

Every circuit-switched network is initially designed for having a sufficient margin to allow communication with low blocking, while some of the traffic resources that are not yet occupied can be utilized for packet data transmission (GPRS). Since the packet traffic has a lower priority than voice traffic, it can be interrupted to accommodate the peaks in circuit-switched traffic.

In Table 3.3 the circuit-switched blocking probabilities for GOS 1% and 2% are calculated, as well as the average number of free TCHs available for GPRS. The number of available TCHs for GPRS traffic would increase as the number of TRX increases.

To accommodate traffic load peaks, some extra resources should be allocated in the system since the distribution of the required data transmissions is not expected to be steady to preserve a minimum user data rate.

3.2.3.2 Defining the Default And Dedicated Territory in GPRS

The default and dedicated territory in GPRS is actually the division of resources between circuit-switched and packet-switched traffic. The default territory is the

Table 3.3 Average GPRS Available Resources

Number of Total TCH	Circuit-Switched Traffic [Erl] (GOS 1%)	Average GPRS Resources [Erl] (GOS 1%)	Circuit-Switched Traffic [Erl] (GoS 2%)	Average GPRS Resources [Erl] (GoS 2%)
7	0.5	6.5	2.9	4.1
14	7.4	6.6	8.2	5.8
22	13.7	8.3	14.9	7.1
30	20.3	9.7	21.9	8.1
38	27.3	10.7	29.2	8.8
46	34.3	11.7	36.5	9.5
54	41.5	12.5	43.9	10.1
62	48.7	13.3	51.5	10.5

number of traffic slots that can be utilized for GPRS purposes; however, if voice calls cannot be accommodated in the system, the default territory is reduced. The traffic slots that are part of the default territory and are only allocated to GPRS usage correspond to the dedicated territory.

To estimate how many slots should be allocated to each of the territories, expected traffic load and traffic profiles should be considered, as shown in Figure 3.4.

If a minimum level of data communication service is required during heavy traffic load situations in the circuit-switched part, at least one traffic slot should be dedicated to GPRS. Clearly, by reserving channels for GPRS, the blocking probability of the circuit-switched part is increasing, since the number of TCH resources available for voice calls is reduced. In Figure 3.5 the blocking probability is depicted as a function of the BTS TRXs, and the number of dedicated TCHs.

Thus, the trade-off between providing a minimum GPRS service level and increasing the blocking probability for the circuit-switched services has to be defined. The priorities of the network operator and the current system performance also play a significant role in this decision.

If the voice traffic that has occupied part of the GPRS default territory increases, this will lead to an automatic reallocation of the TCH resources to the circuit-switched traffic, regardless of the actual GPRS load. As a result, these time slots will be immediately available to users that are willing to establish GPRS sessions or for reallocation of the existing users in order to increase their data rates. If the decreasing circuit-switched traffic is outside the GPRS territory, no automatic reallocation occurs.

3.2.3.3   GRPS Data Throughput

The overall data throughput in a GPRS system depends on a number of parameters. One of the most critical parameters is the CS that is to be applied. As discussed in the previous sections, it depends on the coverage planning and the C/I ratio for the

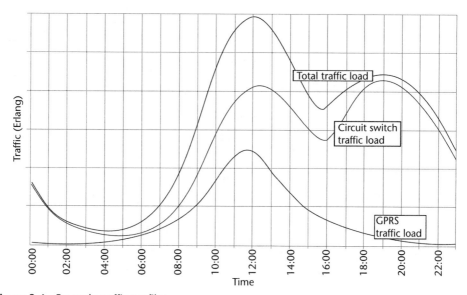

Figure 3.4   Example traffic profiles.

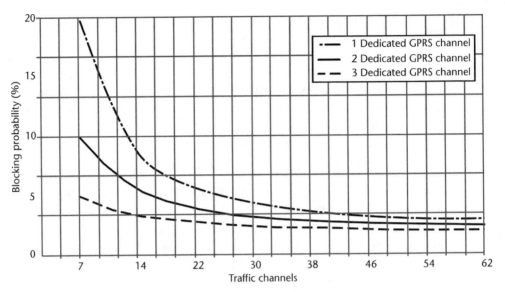

Figure 3.5   Blocking probability versus dedicated channels.

specific area. The theoretical performance of each CS, as a function of C/I, is shown in Figure 3.6.

In the case of an ideal link adaptation the coding scheme changes from CS-1 to CS-2 as the C/I increases from 6 dB to 7 dB, resulting in higher data rates for the users. At this point, it must be mentioned that higher retransmission rates should be expected with higher CSs. This does result in inefficiency for the channel; therefore, a system designed for a C/I of 12 dB will offer higher CS-2 throughput than one designed for 6 dB. In practice, many systems are already designed for C/I levels significantly higher than 6 dB, and hence the regions of interest would be more likely to be the CS-2/CS-3 boundary and the CS-3/CS-4 boundary.

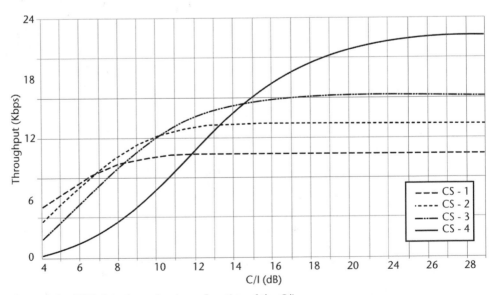

Figure 3.6   GPRS data throughput as a function of the C/I.

The C/I for a given user depends also on his or her location within the area of coverage. Link adaptation is used to apply the CS that will offer a higher throughput considering the blocking error probability. Therefore, it is possible that within a cell there are users attached to GPRS that make use of services with different data rates due to the fact that different CSs are applied for each of them. If, for example during a normal traffic situation, one user requests two time slots for a service and is located very close to the BTS, while another one requesting the same service is located far away from the BTS, then the first user might get the service with double data rates as compared to the second one.

3.2.3.4   QoS in GPRS

As presented in the previous sections, voice is always prioritized, and the only guarantee for GPRS minimum service is the dedicated territory. Subsequently, users often experience data rate reduction due to the increased number of call set up requests. Another factor that can result in this data rate reduction is the increased number of users attached to GPRS. In that case the available bandwidth is split and no QoS can be guaranteed.

The procedure of dimensioning a GRPS network consists of several steps that an operator should follow to configure the network in the best possible way. The first step is the estimation of the traffic profile for the users in the given area. At this point it has to be mentioned, that the traffic profile is strongly linked with the area of coverage, since there are areas where most of the GPRS users are pedestrians or car drivers that make only a little use of the data services, while in other areas (e.g., airports) subscribers may use these services for longer periods of time for downloading files or surfing the Web.

For the network operator it is necessary to consider the QoS while dimensioning the system, since the data rate reduction can be seen as congestion in the packet-switched part of the network. To assess the problem, the mean data rate reduction should be estimated in a given network. Assuming that the circuit-switched traffic is constant and does not enter the GPRS default territory, the packet-switched offered traffic should be estimated based on specific traffic models that can be applied in the area of coverage. Based on this traffic estimation the average data rate reduction can be calculated.

3.2.4   System Parameter Planning

In the previous sections the definition of the territories and traffic dimensioning of the system was presented. Another critical step for the GRPS planning is the configuration of the system parameters, especially the power control–related parameters. Power control is mainly done in the uplink, since network operators rarely use downlink power control.

Power control in the uplink is implemented to improve the spectrum efficiency and to reduce the power consumption in the terminals. The MS follows a flexible power control algorithm that can be optimized by the operator through a number of parameters. In general, GPRS uplink power control is given by the following formula:

$$P_{UL} = \min\left\{\left[P_{C1} - P_{MSmin} - \alpha \cdot (P_{RX} + P_{C2})\right], P_{max}\right\}$$

where

P_{Msmin}: Minimum MS output power, transmitted to the MS message;

P_{C1}, P_{C2}: Constants;

P_{RX}: Received signal strength at the MS;

α: Determines the slope by which the downlink RX_LEVEL affects the power;

P_{max}: Maximum allowed MS transmitting power in the cell.

Both α and P_{MSmin} are cell-specific parameters that are defined by the network operator.

3.2.5 GPRS Performance Monitoring

To efficiently manage the resources of a GPRS network, accurate performance monitoring is required. As presented in the previous chapter, a set of KPIs is used in a GSM system to characterize the system's performance. The KPIs focus on the logical channels, their utilization, and the blocking and drop call rates [11].

In GRPS, the KPIs mainly define the quality level of a connection from the MS to the GGSN; therefore, they focus on the core network, rather than on the physical layer. The GPRS KPIs are classified in the following categories:

- Reliability;
- Delay;
- Peak throughput;
- Mean throughput;
- GPRS utilization;
- Attach attempts failure rate.

3.2.5.1 Reliability

Reliability is probably the most important KPI and evaluates the transmission modes and performance for the different network protocol layers, such as GTP, LLC, and RLC. The GPRS protocol stack that depicts these layers is presented in Figure 3.7. The reliability of a packet-switched session can be considered in two stages, the call set up stage and the release stage. A GPRS session is characterized successful, if both the call set up and call release are completed successfully. In the event that one of these is not successful, the blocking of GPRS resource requests can be calculated by the number of resource requests and resource allocations.

The GPRS blocking of resources is linked with the temporary block flow (TBF) set up failure, due to the lack of PDTCH resources. TBF is the physical connection that is used to support the unidirectional transfer of LLC PDU on packet data physical channels. TBF is allocated in one or more TCH channels and comprises a number of RLC/MAC blocks carrying one or more LLC PDUs. A TBF is temporary as its name denotes and is maintained as long as data is transferred. PDTCH access failure

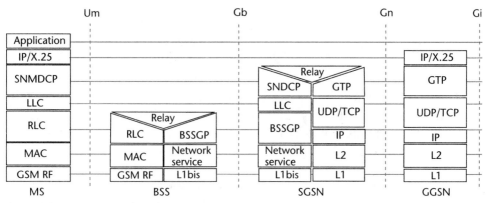

Figure 3.7 GPRS protocol stack.

can be separated in two types. The first type is called hard blocking, and the second one is called soft blocking.

Hard GPRS blocking.   Hard blocking describes the situation when no radio resources are assigned after the GPRS service request. For instance, if we consider the case when a user requests three time slots (TSLs) to download a file using ftp and no resources at all are assigned, this is characterized as hard GPRS blocking.

Soft GPRS blocking.   Network operators usually operate their GPRS networks using at least one dedicated channel. For that reason hard blocking occurs rarely, and a more common situation is the soft GPRS blocking where users are assigned a portion of the resources they request. For example a user who requests three time slots for a service will finally receive one or two.

The reliability KPI can also be divided into the session release success rate and data connection success rate indicators. They are described in Sections 3.2.5.1.1 and 3.2.5.1.2.

3.2.5.1.1   Session Release Success Rate

This KPI discriminates the various reasons for abnormal TBF release expressing the percentage of TBF dropped sessions due to bad radio conditions as well as due to high circuit-switched traffic, where the prioritized voice requests are served by assigning traffic slots from the default GPRS territory. The session release success rate indicates the impact of circuit-switched (CS) traffic on packet-switched (PS) traffic (TBF drops) when radio resources (TCHs) are not sufficient, and CS accommodated traffic is served also by using traffic slots from the default GPRS territory. Hence, when experiencing a high drop ratio because of this reason, dedicated GPRS capacity is needed to assure a minimum QoS. The following formulas calculate the GPRS uplink (UL) and downlink (DL) dropped TBF ratio:

GPRS Dropped UL TBF Ratio (%) =

$$= \frac{sum(\text{UL release due to CS traffic} + \text{UL release due to no response from MS})}{sum(\text{Number of TBF})} \cdot 100$$

GPRS Dropped DL TBF Ratio (%) =

$$= \frac{sum(\text{DL release due to CS traffic + DL release due to no response from MS})}{sum(\text{Number of TBF})} \cdot 100$$

where UL/DL release due to CS traffic indicates the number of UL/DL TBF releases due to the CS traffic. It is updated when the circuit-switched traffic builds up to such an extent that the GPRS TBFs have to be released to allow the CS traffic to seize the GPRS default TCH.

In the same situation, UL/DL release due to no response from MS indicates the number of UL/DL releases due to no response from the corresponding MS. This occurs when there is sufficient interruption in the data transfer for the packet control unit (PCU) to consider the MS as lost. The time associated with this counter depends on whether the transfer is in the downlink or in the uplink, on the kind of activity on the time slot, the MS multislot class allocation, and the BLER. In most of the cases the timing can range from 300 ms to a few seconds.

3.2.5.1.2 Data Connection Success Rate

Whenever a mobile station is admitted to access the GPRS channels and requests a service, the network will establish a connection that supports the unidirectional transfer of LLC PDU. This connection is the TBF one that was described in the previous section. There are three TBF types when a mobile station is in data transfer:

- Uplink TBF;
- Downlink TBF;
- Simultaneous uplink and downlink TBF.

The TBF is assigned to the radio resources of one or more PDTCH and comprises a number of RLC/MAC blocks carrying one or more LLC PDU. The TBF set up failure due to the lack of PDTCH resources results in soft or hard blocking in GPRS. The following formulas present the uplink and downlink hard blocking rate in PDTCH.

$$\text{Hard UL PDTCH Blocking Rate (\%)} = \frac{sum(\text{Number of Radio resources UL TBF})}{\sum_{1}^{4} n \cdot n \text{ channels request in UL}} \cdot 100$$

$$\text{Hard DL PDTCH Blocking Rate (\%)} = \frac{sum(\text{Number of Radio resources DL TBF})}{\sum_{1}^{4} n \cdot n \text{ channels request in UL}} \cdot 100$$

where number of radio resources UL/DL TBF is a counter that is updated when a new uplink TBF is requested by the MS, but the PCU has no radio resources available for the uplink TBF. It is also updated by the failure of reallocating the existing uplink TBF, due to the lack of PDTCH resources. That means that all the available PDTCH channels are occupied.

All of the above formulas do not take into account signaling, which is considered separately and can be found from performance indicators characterizing the RACH and AGCH performance. If the statistical analysis shows that there is blocking, the most possible reason is that the territory has been smaller than the default setting defined from the corresponding CS used. This is because the CS side will return the default channels back to the PS territory as soon as the CS load allows it.

The formulas below calculate the uplink and downlink soft blocking rate in PDTCH:

$$\text{Soft UL PDTCH Blocking Rate (\%)} = 100 - \frac{\text{Total allocated TSL}}{\text{Total requested TSL}} \cdot 100 = 100 -$$

$$-\frac{\sum_1^4 n \cdot n \text{ channel allocation in UL}}{\sum_1^4 n \cdot n \text{ channel request in UL}} \cdot 100$$

$$\text{Soft DL PDTCH Blocking Rate (\%)} = 100 - \frac{\text{Total allocated TSL}}{\text{Total requested TSL}} \cdot 100 = 100 -$$

$$-\frac{\sum_1^4 n \cdot n \text{ channel allocation in DL}}{\sum_1^4 n \cdot n \text{ channel request in DL}} \cdot 100$$

The formulas presented above refer to mobile stations requesting in UL or DL no more than four traffic slots.

3.2.5.2   Delay

The *delay* is another critical KPI that does not occur due to congestion in the radio resources, but it appears and is considered in the core network. It specifies the end-to-end transfer delay measured in the transmission of SDUs through the GPRS network. This KPI has direct interaction with the use of the CCCH of the CS channels, such as RACH, AGCH, and PCH (which are called PRACH, PAGCH, and PPCH when they are used in GPRS). A big delay is strongly linked with congestion in the above resources.

CCCH signaling is used for paging and for uplink and downlink TBF connection set up. GPRS paging is made on the PCH. The MS initiates uplink TBF establishment on the RACH. The network responds to the MS on the AGCH. The network initiates a TBF on either the AGCH or the PAGCH depending on whether a TBF is already established. The uplink and downlink transfer will require extensive use of the RACH and AGCH channels owing to the number of TBFs required for supporting the higher level protocols and their interworking with the application.

The load on the control channel is highly dependent on the service that is requested. The average rate at which the AGCH messages are sent is about twice the rate at which the RACH request messages are sent. Since there is more capacity on RACH than on the AGCH, the limiting factor in signaling is the AGCH. In a combined control channel configuration AGCH provides approximately seven control messages per second, which can result in up to a 30-Kbps connection. However, it should be noted that this is the average requirement on the AGCH and not the peak one. The peak requirement exceeds the AGCH capacity with an offered load of approximately 11 Kbps, consequently offering higher GPRS loads, which will cause blocking in the AGCH, which in turn will result in delays for the packet traffic and an increased set up failure rate for the circuit-switched user. In addition to the load originating from TBF establishment, mobility management procedures such as routing area updates and cell reselections as well as GPRS attach and detach generate common control signaling load. In Figures 3.8 and 3.9, the expected throughput and delay for the allocation of one and two time slots are depicted.

3.2.5.3 Average User Data Throughput

The average user data throughput is the KPI that shows the average rate at which the data is being transferred in the network for the PDP context. This important KPI is the average throughput offered to the users in the air interface. Actually, the most important KPI concerning the performance monitoring of the GPRS network is the offered throughput per user. The user discrimination can be done only on a TBF session level. One user can request many TBFs just to download for example a simple Web page or to send an e-mail. But, as it was described before, it is very difficult to measure the offered throughput per TBF by the network. An alternative way of doing this is the measurement of offered throughput per TS that gives something similar but not from the user point of view but from the network point of view. For example the target for the network is to provide 13.4 Kbps per time slot, for CS-2. Actually, this throughput level also includes the RLC and MAC layer headers that are not considered in the related formulas to calculate the offered bit rate being perceived as the closest possible to the user's point of view. If we disregard this factor, the actual effective throughput per TSL in the LLC/RLC layers is around 12 Kbps, a value that should be measured in the air interface if no retransmissions take place. In particular, the throughput calculation on the LLC layer is the last layer where the TBF and time slot distinction can be performed.

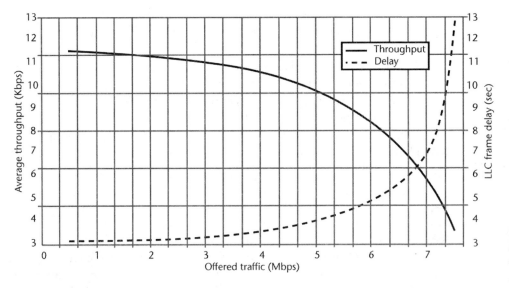

Figure 3.8 Throughput and delay for 1 RTSL allocation.

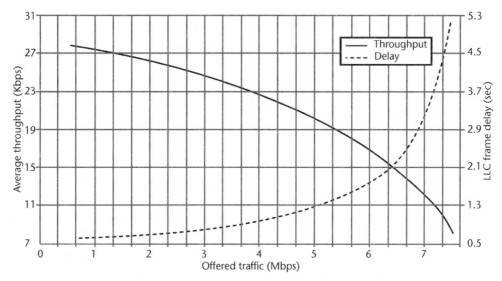

Figure 3.9   Throughput and delay for 2 RTSLs allocation.

Of course, in case there is more than one user requesting GPRS service in this time slot they will share the overall maximum throughput that can be given by this time slot. However, the measured throughput will be mainly related to the radio conditions since possible errors in the transmitted blocks will eventually decrease it.

Before the definition of the throughput formulas, some clarifications regarding the counters used in traffic calculation is needed. The KPIs mentioned until now provide the useful data rate per time slot, since for the moment the data rate per connection cannot be given. The numerator includes the user data (RLC data blocks in kilobits per second), while the denominator includes all the blocks transmitted, divided by 50 (maximum number of blocks per second). In addition to the radio blocks consisting of four bursts there are also some idle frames in between on a PDTCH. These idle frames are needed for neighbor cell BSIC decoding and for the continuous timing advance procedure. In a 52-multiframe on a PDTCH there are 12 radio blocks and four idle frames, $12 \cdot 4 + 4 = 52$ TDMA frames. One 52-multiframe has duration of $52 \cdot 4.615$ ms $= 240$ ms. Thus, on average there are 12 blocks/240 ms $= 50$ blocks/s. The formulas below give the average data throughput offered per timeslot.

$$\text{Average effective UL throughout per used TSL } = \frac{\text{UL payload in (kbytes)}}{\text{UL time for data transfer (sec)}}$$

$$= \frac{\text{data in kilobits}}{\text{TBF total time}}$$

$$\text{Average effective DL throughout per used TSL } = \frac{\text{DL payload in (kbytes)}}{\text{DL time for data transfer (sec)}}$$

$$= \frac{\text{data in kilobits}}{\text{TBF total time}}$$

3.2.5.4   Peak User Data Throughput

The *peak* throughput is the maximum rate at which the data is being transferred in the network for the specific PDP context used (e.g., IP and X.25). This important KPI shows the availability of the PDTCH. This KPI is being referred for one mobile user and is given after the processing of the reports generated for each user.

3.2.5.5   GPRS Utilization

Another very useful KPI, especially in this early working phase of the GPRS network, is the monitoring of usage of the related resources as well as the total amount of data transmitted in a specific period of time. It indicates how many time slots the GPRS traffic consumes on average during this period. This information is useful, for example, in forecasting the need for applying capacity extensions, and it represents Erlangs on PDTCH. The idea behind this indicator is the amount of GPRS territory resources usage measured in Erlangs or TSLs, meaning how many TSLs were occupied during a specific period of time in order for a known amount of data to be transferred.

The allocation of physical channels for GPRS usage can be done for transferring either normal data or signaling data. The calculation of the traffic is made on the basis of the number of RLC layer blocks transmitted in a specific period of time based on the data that can be sent in one TSL for the same period and is given by the following formulas.

$$\text{UL GPRS timeslot usage }(\%) = \frac{\text{Actual UL data throughput (blocks)}}{\text{max. nbr of blocks during measurement period}}$$

$$= 100 \cdot \frac{\text{Data blocks tansmitted \# greater one chosen UL}}{(\text{available GPRS channel time in sec}) \cdot (\text{nbr of blocks per sec})}$$

$$\text{DL GPRS timeslot usage }(\%) = \frac{\text{Actual DL data throughput (blocks)}}{\text{max. nbr of blocks during measurement period}}$$

$$= 100 \cdot \frac{\text{Data blocks tansmitted \# greater one chosen DL}}{(\text{available GPRS channel time in sec}) \cdot (\text{nbr of blocks per sec})}$$

Finally, the *total amount of data volume* indicates the total amount of data transmitted as CS-1 or CS-2 blocks, in the uplink or downlink. MAC and RLC header bytes are not included in order to get as close as possible to the payload data coming from the IP/LLC layers.

3.2.5.6   Attach Attempts Failure Rate

As it is known, whenever a GPRS subscriber leaves the *idle* state and enters the *standby* state, the MS has to *attach* to the network, thus providing information about its current location. Then if data are sent or received the subscriber is considered to be in the active state. The three states in which a GPRS subscriber may be are depicted in Figure 3.10. The attach procedure is similar to the location update in GSM.

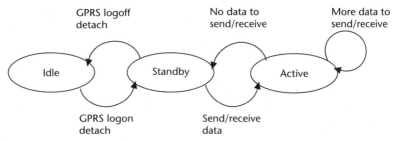

GPRS logoff   No data to   More data to
detach   send/receive   send/receive

Idle   Standby   Active

GPRS logon   Send/receive
detach   data

Figure 3.10   GPRS states.

Failure to attach to the network can be calculated as follows:

GPRS Attach Attempts Failure Ratio (%) =

$$= 100 \cdot \frac{\text{(Failed GPRS attachments)}}{\text{Succesful GPRS attachments + Failed GPRS attachments}}$$

The above KPI could be part of a wider scope KPI concerning attachments in a GPRS network, which are divided into the following three categories:

1. *Attach for GPRS services only:* In a GPRS attach, the subscriber registers to the network for GPRS services only. This means that the subscriber wants to use the data services of the GPRS network and not its CS services.
2. *Attach to both GPRS and CS services:* In a combined attach, the subscriber registers to the network to use both GPRS and CS services. This means that the subscriber wants to use both the data services of the GPRS network as well as its CS services.
3. *Attach to GPRS services, when already CS-attached:* In an IMSI already attached, the subscriber is already attached to CS services and intends to use the GPRS services of the network.

3.3   Resource Management Techniques and Guidelines for Implementation

This section presents resource management techniques for 2+G cellular systems as well as guidelines for implementation in a real networking environment. The first four techniques presented here are similar to those presented in Chapter 2, but this time their effect is measurable in the packet-switched part of the network.

Finally, a number of techniques that aim to increase the performance of GPRS by focusing on packet scheduling are presented. The proposed resource management techniques have the significant advantage that they are algorithms that can easily be implemented with a programming language. The complicated adaptation of the implemented software on the GPRS architecture and the difficult access to the related network interfaces are the techniques' main disadvantages. The proposed algorithms are performed on the LLC frame level and the RLC block level. Unfortunately it is very difficult for the operator to change the configuration of these interfaces. If operators want to change their configuration they have to add specific extra modules to the current GPRS architecture.

Packet scheduling is performed on two different levels. These levels are the LLC frame level and the RLC block level. The transmitter obtains the channel quality information from the radio link control (RLC)/medium access control (MAC) acknowledgment messages [5] that contain information regarding the successfully received radio blocks. Scheduling is performed at the radio blocks level in the RLC/MAC layer, and it is assumed that this layer operates in acknowledge mode with a maximum of three retransmissions. On the other hand, the LLC layer operates in nonacknowledge mode with a maximum payload size of 1,520 bytes. Scheduling on the LLC frame level implies that the queue order can be changed on every LLC frame (maximum 1.6 kbytes) while scheduling on the RLC block level allows changes every 20 ms. These interfaces are the only interfaces that the packet scheduling algorithms affect. In fact, the RLC and the LLC interfaces are inaccessible to operators and if they want to change their configuration they have to add specific modules as we have mentioned already. For that reason the proposed packet-scheduling algorithms cannot be implemented on an existing GPRS network, but they could be implemented in a future GPRS version installation.

3.3.1 Dynamic Cell Resizing with the Use of C2 Values

It is important to note that the GPRS service uses the same C2 value as in the circuit-switched services. This technique is described in Section 2.5.5 for standard GSM systems. It is based upon the assumption that the mobile station should camp within the cell offering the best coverage. Also the cell reselection based on the C2 value in the GPRS is the mechanism corresponding to handover in the circuit-switched services. This method performs best when it is used with prediction.

The purpose of C2-based cell reselection is to control the cell reselection of the MSs in idle mode. The C2-based cell reselection allows the operator to define extra criteria for reselection in addition to the power level and to define different reselection criteria for different kinds of cells—for example, for macrocells or microcells.

The C2 reselection information is broadcasted to the MSs in system information 3 and 4 frames. The BSC sends the system information frames to the BTS on the A_{bis} interfaces, and the BTS sends them on the air. The MS evaluates the C2 reselection criteria according to the given parameters. The C2 reselection criteria also include the so-called penalty time, which allows the cell reselection to happen more slowly than the pure C1 criteria evaluated by the MS. This enables avoiding undesirable reselection of a microcell in an environment where there is coverage by a microcell and a macrocell.

The packet-switched services are using the PBCCH, PCCCH, and PTCH resources, which are similar in nature and functionality with the BCCH, CCCH, and TCH. Thus, the implementation of the technique has the same effects in the GPRS services as in the standard circuit switched ones. This resource management technique is appropriate for GPRS congestion, especially when there is a significant demand for common resources like PAGCH and PPCH, the nonavailability of which degrades the quality of both circuit- and packet-switched services.

This method does not affect any network elements, but it does affect network planning. In situations where the reduction of the coverage area of a congested cell is unavoidable, a decrease in the C/I ratio of about 5 dB is also expected, in worst cases. This may lead to a reduction in the code scheme used in GPRS service, for example from CS-2 to CS-1 and from 13.4 Kbps to 11.5 Kbps data rates. In practice the overall gain is much higher than this reduction. Also, for cells that are located at the border of two different LAs, different settings are selected for cell reselection hysteresis, to prevent useless LUs as we did for standard GSM networks.

The parameters involved in this method are the same as for the standard GSM implementation presented in Section 2.5.5.1. To activate the C2 values, the network topology should be planned with overlapping areas between cells. GPRS service can work with CS-4 and without hopping if the normal planning is implemented with a C/I of 17 dB. In general the implementation of the method is the same as in the standard GSM of phases 1 and 2 (see Section 2.5.5).

3.3.2   Adjusting the Rx Level

This method is similar to the one for standard GSM systems (phase 1 and 2) as presented in Section 2.5.10. It preempts a number of GPRS users from using the network with the only criterion being their location in the cell. As with the C2 values method this technique has an impact on the cell reselection mechanism of the GPRS terminal that is very important for the QoS that the user will experience.

3.3.3   Modifying the BCCH Frequency List Due to Traffic Congestion

This method is similar to the one for standard GSM systems (phase 1 and 2) as discussed in Section 2.5.11. When a cell is congested, the operator can decrease the offered telecommunication traffic by reducing the number of the new incoming MSs, which tend to camp in the congested cell. An efficient way to do this is by modifying dynamically the BCCH frequency lists that are transmitted from its neighbor cells. In that way the congested cell is going to be excluded from the new list and so this problematic cell will not be a candidate for reselection or handover in both GPRS and circuit-switched services. As a result the offered traffic for the congested cell will be decreased, and it will be distributed to other nearby cells.

3.3.4   Exploiting the Radio Resources of Adjacent Cells

A system designed only for circuit-switched traffic will allow a basic GPRS throughput since it usually has sufficient margins in the available resources for achieving a low blocking level, and thus some of the spare instantaneous capacity can be utilized for packet data transmission. As long as the packet traffic can be temporarily interrupted to accommodate the peaks in circuit-switched traffic, no degradation in the circuit-switched services will occur.

As described in the previous sections, a number of slots are characterized as GPRS default territory, while only the dedicated GPRS slot(s) will remain for packet-switched communication, in the case of increasing voice call requests. An important point is that the default slots should be adjacent. If, for example, the eight slots of a TRX are indicated with *slot 0* to *slot 7*, in the case that four slots are the default to be used for GPRS, *slot 4* to *slot 7* can be allocated. Assuming now that there are four ongoing calls and that there is a request for a new call establishment, this will be assigned to *slot 4*, and the throughput of the data communication will be decreased. A solution that would avoid this would be to apply forced handover to the voice user that occupies *slot 3* well in advance, so that the new incoming call could be accommodated in that slot, instead of limiting the GPRS bandwidth. This technique can be easily realized, since most of the manufacturers can execute the forced handover command for a specific slot of the TRX.

The main advantage of this technique is that QoS in GPRS service is achieved without increasing the blocking probability of the voice call requests; users may just experience a slight interruption during handovers.

3.3.5 Modified Earliest Deadline

GPRS should satisfy the QoS requirements posed by modern data communication applications and for that reason specific packet scheduling strategies have been devised and deployed. On the other hand, GPRS introduces four coding schemes (CS-1, CS-2, CS-3, and CS-4) with different degrees for data protection according to the radio channel conditions. Then, the selection of the appropriate coding scheme is done in accordance with the time-variant channel conditions in order to achieve the best possible QoS at any time.

We now discuss the modified earliest deadline (MED) scheduling policy, using the link adaptation strategy, based on BLER estimation, and heterogeneous traffic is presented. This technique has been used for asynchronous transfer mode (ATM) networks and has been adapted for GPRS [11–13]. The performance of the proposed scheduling policy has also been evaluated in terms of throughput and delay. For urban environments with high load conditions, the obtained results show that a simple link adaptation strategy enhances the system's performance, even though there is a trade-off with respect to the packet loss in the network [14, 15].

3.3.5.1 Parameters

This section presents and analyzes the most important parameters related to this technique. These parameters have a very crucial role on the execution of the proposed packet-scheduling algorithm.

The GSM BSS is responsible for sharing the radio resources among both circuit- and packet-switched services. Therefore, for the packet management through the radio channel the BSC must also incorporate the PCU. The physical channels, available in a cell, are dynamically shared between GPRS and GSM services. The channels that are associated with GPRS are called PDCHs, and the basic transmission unit of a PDCH is called a radio block. To transmit a radio block a TS in four consecutive TDMA frames is used. A PDCH is structured in multiframes comprising of 52 TDMA frames, which correspond to a duration of 240 ms. The mean transmission time per radio block is 20 ms. A radio block contains 456 bits (114 per burst). The structure and the number of payload bits of a radio block depend on the message type and coding scheme as presented in Table 3.4.

Table 3.4   Coding Parameters for the Coding Schemes

Channel Coding Scheme	Data Bits in Radio Block	Data Rate per Time Slot (Kbps)
CS-1	181	9.05
CS-2	268	13.4
CS-3	312	15.6
CS-4	428	21.4

To use the scarce radio resources in a more efficient way and to support a number of applications with different requirements, GPRS provides several QoS profiles. These QoS profiles allow the mobile operators to create schemes for charging differentiation. In the GSM 2+ phase, the QoS profiles are characterized by the five parameters [service precedence (priority), reliability, delay, mean user data rate, and maximum user data rate] that are listed in Table 3.5.

The radio channel quality conditions may vary during the connection between the MS and the BSS. The goal of the link adaptation algorithm is the selection of the best coding schemes according to channel conditions in order to reach a throughput as high as possible. The proposed link adaptation algorithm is based on the use of the BLER parameter as the channel quality indicator. This parameter is evaluated at the end of every message of the acknowledgment window. The BLER estimation is calculated using the following expression:

$$\overline{BLER(k)} = \overline{BLER(k-1)} \times (1-\alpha) + BLER(k) \times \alpha$$

where $\alpha = 0.8$ and $k \in [1, \text{window Ack size}]$.

This expression is equivalent to a first order finite impulse response (FIR) filter, and it is used to weigh the samples, being the weight of the most recent RLC/MAC data blocks higher. The α value controls the memory of the algorithm.

Once the BLER is estimated, the equivalent committed information rate (CIR) is calculated according to the current coding scheme. Finally, the link adaptation algorithm moves to the finest coding scheme according to the Table 3.6 criteria. The new coding scheme is applied on the next logical link control (LLC) frames. Note that this link adaptation algorithm is channel dependent.

There are several different approaches that can be used to determine the initial coding scheme. In the downlink, the best coding scheme is chosen according to the distance from the BTS. It is proved that for the urban case, with a cell radius (Rc) of

Table 3.5   The QoS Profile

Parameter	Values					
Precedence	Hight,normal, low					
Reliability	Packet loss probability; duplicate, out of sequence					
Delay	Size SDU	Class	1	2	3	4
	128 (bytes)	(Means)	< 0,5	< 5	< 50	Best effort
		95%(s)	< 1,5	< 25	< 250	Best effort
	1,024 (bytes)	Mean(s)	< 2	< 15	< 75	Best effort
		95%(s)	< 7	< 75	< 375	Best effort
Rate	Mean	Depend				
	Maximum					

Table 3.6   C/I Thresholds

Finest Coding Scheme	*CIR Threshold*
CS1	CIR <= 7.5 dB
CS2	7.5 dB < CIR <= 10 dB
CS3	10 dB < CIR <= 21 dB
CS4	CIR > 21 dB

3 km and a frequency reuse parameter (K) equal to 4, the CS4 scheme is optimal for distances lower than 2,010m while the CS-3 scheme performs better for the interval 2,010m < Rc ≤ 3,000m. The link adaptation is implemented only in the downlink direction.

3.3.5.2   Expected Results

The performance of the proposed scheduling and link adaptation algorithm was examined using a GPRS software simulator implemented in MATLAB. To generate the different combinations of data traffic, three types of traffic sources were considered, namely: e-mail, WWW-wireless application protocol (WAP), and short message applications.

The model used for e-mail application considers that incoming user messages were stored in a dedicated r-mail server. In our simulation, a comprehensive model for Web traffic, called the "behavioral model" was adapted modifying the mean value of the IP packet length to emulate the comportment of the WAP in the context of a GPRS system. Finally, the SMS, which is one of the most popular services in the actual GSM system, was also included in our simulation. For SMS traffic, the length of the unit short message was limited to 140 bytes in order to take into account the restrictions in the mobile application part (MAP) signaling layer. For all the models, the packet or session interarrival time distribution was assumed to be exponentially distributed. In our analysis, it was assumed that there were two TSs exclusively dedicated for the GPRS system in the cell. No IP header compression was used within the subnetwork dependent convergence protocol (SNDCP) [1]. Service scheduling is performed at the radio blocks level in the RLC/MAC, and it was assumed that this layer operated in acknowledge mode with a maximum number of retransmissions equal to three. The maximum number of simultaneous active mobile stations in the downlink direction was 32 and 15 in the uplink case. All the MSs were randomly distributed in the radio cell $(R = 3$ km).

In our simulation when a fixed CS-1 coding scheme was assumed, the link utilization could grow up to 70% still achieving the QoS delay requirements. However, when link adaptation was used, the link utilization increased substantially, being the maximum number of simultaneous active MS the true system's limitation. The link utilization was defined as the degree of occupation of available resources for an hour, (i.e., if all resources were occupied for an hour then the link utilization was 1).

During the test the evolution of the throughput as function of total number of sessions/hour for the proposed algorithm is also observed. The throughput is defined as the amount of error-free user data that reaches the destination during the simulation time. That happened because the possible retransmissions at the RLC/MAC level were not taken into consideration. From the results, one could see that for e-mail applications it was possible to increase the system throughput up to 110%, with respect to the use of a fixed CS-1 coding scheme, when link adaptation was assumed.

The evolution of the packet loss probability versus the total number of sessions/hour when link adaptation is used was also considered. The results showed that packet loss was approximately constant around 4% with LA and very low (much less than 1%) for the CS-1 fixed coding scheme. As a conclusion we can point out that for LA, there is a trade-off between the throughput increase and the loss probability.

3.3.6   Minimum Laxity Threshold

This section introduces another scheduling policy, namely the minimum laxity threshold (MLT), which uses the link adaptation strategy based on BLER estimation; the section also introduces heterogeneous traffic. The performance of the proposed scheduling policy has also been evaluated in terms of throughput and delay as we have done with MED. For urban environments with heavy load conditions, the obtained results show that a simple LA enhances the system's performance, even though there is a trade-off with respect to the network packet loss as it was the case also with MED.

The MTL policy is based on the laxity concept [16]. The laxity, in the case of GPRS, represents the amount of time for which the PCU scheduler may remain idle, or serving radio blocks of other classes, and still it is able to transmit a given radio block before the expiration of its deadline [15]. It is assumed that the radio blocks in the buffer are sorted according to their deadlines.

Denote as RBk the number of radio blocks stored in the class k queue and $dk(j)$ the deadline of the j-th radio block of this class k. The laxity $Lx1(j)$ of the jth radio block allocated in the queue with the highest priority (Q1) at the time t is defined by:

$$L_{x1}(j) = d_1(j) - t - \left[(j-1) \cdot tx(RB)\right]$$

where $tx(RB)$ is the radio block transmission time (~20 ms). The laxity for radio blocks allocated in a queue (Q2) that has lower priority than the previous one is given by:

$$L_{x2}(j) = d_2(j) - t - \left[RB_1 \cdot tx(RB)\right] - \left[(j-1) \cdot tx(RB)\right]$$

The above expression takes into account that there are RB1 radio blocks allocated in the queue Q1 that shall be served. A similar expression can be obtained for the Q3 priority queue. Prior to each radio block transmission time, the laxities are evaluated for each of the radio blocks in the queues of traffic classes and the minimum laxities are computed for the queues themselves, that is:

$$L_k = \min(L_k(i))$$
$$0 \le i \le RBi$$

A threshold operation is then used to choose which QoS to serve. If $Lx1 < 2 \cdot tx$ (RB) then the high-priority queue (Q1) must be served. If not, the minimum laxity of the second-priority queue is compared. If this second priority queue cannot be served, then the queue with the third priority class is considered. If neither of these conditions is true, then best effort traffic can be transmitted.

3.3.6.1 Parameters

The MLT method uses the same parameters as MED that are presented in Section 3.3.5.2. It uses the link adaptation algorithm that is based on the estimation of the BLER parameter.

3.3.6.2 Expected Results

The performance of the proposed scheduling and link adaptation algorithm was examined using the GPRS software simulator. This simulator is exactly the same with the simulator for the estimation of the performance of the MED algorithm, and the same traffic sources and initial configuration are used. In this paragraph we present the results of the simulation for the MLT algorithm.

In our simulation when a fixed CS-1 coding scheme was assumed, the link utilization could grow up to 70% still achieving the QoS delay requirements. However, when link adaptation was used, the link utilization increased substantially, with the maximum number of simultaneous active MSs the true system limitation. The link utilization was defined as the degree of occupation of the available resources for an hour (i.e., if all resources were occupied for an hour then the link utilization was 1).

In the uplink case, for medium and heavy link utilization (higher than 50%), the normalized delay was higher than 1 (the system falls in congestion) for both QoS 1 and QoS 2 service classes. This was due to the fact that in the uplink case, the multislot capability was limited to 1 TS, and therefore the packets remained in the queue longer time before they were transmitted.

Another QoS parameter studied in our simulation was the maximum delay observed in the 95% of the cases. From the results it can be appreciated that this parameter is less restrictive than the mean delay because, for all the services, the normalized delay was less than 1 for both the fixed coding scheme (CS-1) and link adaptation strategies.

To have a complete vision of the behavior of the proposed strategy, it was also important to analyze the mean waiting time (delay 1 RB) before the transmission of the first radio block of a packet. From the simulation results ones can see that, for the MLT algorithm the mean waiting time was kept low when the traffic load increased.

Finally, for each traffic pattern and for the proposed scheduling strategy, Table 3.7 shows the maximum number of the sessions/hour that can be admitted to cope with the requirements in terms of quality of service. It is also true that for LA, the parameters that limit the allowed number of sessions are the link utilization as well as the maximum number of simultaneous users.

Table 3.7   Comparative Sessions/Hour for MLT

Discipline	MLT	
Traffic	λ (CS1)	$\wedge$ (LA)
WWW	< 70	< 190
E-mail	< 150	< 330

3.3.7   Virtual Clock

In this paragraph, we introduce the virtual clock (VC) scheduling strategy using heterogeneous traffic [17]. The performance of the proposed scheduling policy has also been evaluated in terms of throughput and delay. The simulation results showed that the VC scheduling technique is very fair, and it performs well enough in the context of a GPRS system with heterogeneous traffic, even under heavy network load conditions. Moreover, the implementation complexity is low.

The VC strategy aims to emulate the behavior of an ideal time division multiplexing (TDM) system. For each packet a virtual time stamp is assigned. This virtual time stamp indicates the time that the packet would be transmitted assuming an ideal TDM strategy. The packets are transmitted in increasing order of their virtual time stamp. The VC algorithm queues the incoming packets according to their virtual time stamp, called finish number. Every flow has its own finish number. The finish number is determined using the negotiated end-to-end transmission rate, taking into account both the needs of applications and the available resources. Each flow has its own clock that advances for every new incoming packet in the flow. At specified time intervals, the algorithm compares the virtual clock of the flows with the real system clock. If the virtual clock is faster, it means that the flow is sending too fast. In that case, the flow penalizes itself. The packets of this flow are queued at the end of the transmission queue, and they would be eventually dropped. When adapting the algorithm for GPRS, the finish number is calculated on a radio block basis.

3.3.7.1   Parameters

The VC method uses the same parameters as MED presented in Section 3.3.5.2. In addition, as we have seen in Section 3.1.2 the PDP context specified in GPRS is defined as the information sets, allocated in the MS, SGSN, and GGSN nodes, which are used to specify the tight relationship between an application of a mobile subscriber, a PDP type, and one QoS profile. Several PDP contexts with different QoS parameters can share the same PDP address (secondary PDP context). To define a QoS contract between the MS and the network, PDP contexts containing QoS profiles are negotiated between the MS and the SGSN node. With the introduction of the QoS concept, it is possible to use the network resources in a more efficient way. That is, the application data flows are managed according to their actual needs.

Even though, from the application viewpoint the QoS is end-to-end defined, only the QoS issues related to the radio network segment are considered, because they represent the bottleneck of a typical GPRS data connection. The QoS profiles are characterized by the five parameters listed in Table 3.5.

3.3.7.2   Expected Results

The performance of the proposed scheduling and link adaptation algorithm was examined using the GPRS software simulator. This simulator is the same with the simulator for the estimation of the performance of the MED algorithm, and the same traffic sources and initial configuration are used. In this paragraph the results of the simulation for the VC algorithm are presented.

In our experiment only the downlink performance was analyzed. Table 3.8 depicts the values for the QoS and traffic speed.

The assigned flow bandwidth depends on the type of the assumed QoS. The total link output bandwidth was distributed among all different classes. The negotiated rate for class I was 50% of the available bandwidth, while for class II it was 30%, and the rest was for class III and best effort. Note that for the VC algorithm the QoS profile was guaranteed, because it could achieve up to the 90% of link utilization while meeting the delay requirements. When link utilization is increased, the VC scheduling algorithm took in consideration the history of all flows. Therefore, after a long inactivity period, when a new packet arrived at the flow it was served first. When the network traffic increased, QoS_2 was obviously penalized and for VC strategy this event appeared earlier. In summary, for the simulated environment, it could be concluded that the main system limitation was the maximum number of simultaneous temporary flow identities (TFIs) [5] available rather than the packet delays, even for heavy load traffic conditions.

For the VC strategy, the evolution of the throughput versus the total number of sessions/hour is also observed. In this observation the throughput was defined for each user packet as the number of bits divided by the transmission time (queuing time + air interface transmission). At the end of simulation the mean throughput was calculated for all application packets. Besides, the possible retransmissions at the RLC/MAC level were not taken into consideration. From the results, it can be concluded that the VC algorithm provides a soft throughput degradation when the traffic load is increased. This was due to the fact that every connection had its independent finish number, and, therefore, a packet from a given flow was served if it had the lowest finish number. This also explains the better throughput of the e-mail service because the e-mail packets were smaller than the Web ones.

During the simulation, the behavior of the VC strategy was also estimated in terms of the mean waiting time to transfer the first radio block, assuming QoS_1. We noticed that the delay for the VC strategy was too high. However, when a new packet of a given flow arrived to the PCU, the time stamp, which finally fixed the delay for transmitting the first radio block, became affected by previous transmissions of this flow. To obtain a complete characterization of the performance of the proposed strategy, the 95% delay, as defined by ETSI, was also analyzed in this simulation.

Finally, for each pattern of traffic and for the VC strategy, the maximum number of the sessions/hour admissible meeting QoS delay is reported in Table 3.9.

Table 3.8   Simulation Scenario

Services	Direction	Delay Class	Portion
WWW	Downlink	QoS 2	20%
E-Mail	Downlink	QoS 1	27%
SMS	Downlink	QoS 2	13%

Table 3.9 Sessions/Hour for VC Strategy

Discipline	VC
Traffic	λ (CS1)
WWW	< 120
E-mail	< 160
SMS	< 80

3.3.8 Modified Advanced Time

This section introduces the modified advanced time (MAT) scheduling strategy using heterogeneous traffic. The performance of the proposed scheduling technique has also been evaluated in terms of throughput and delay. The simulation results showed that the MAT as a resource management technique is very fair and it performs well enough in the context of a GPRS system with heterogeneous traffic, even under heavy network load conditions. Moreover, the implementation complexity is low [18]. This algorithm was introduced in [19], and an adaptation for GPRS was presented in [20]. For each data flow, this scheduling strategy establishes the decision to transmit a given data block on the use of an index [$Iu(t)$] that reflects whether the flow has overused or underused the assigned radio resources.

3.3.8.1 Parameters

The MAT RRM technique uses the same parameters with the MED technique as presented in Section 3.3.5.2.

3.3.8.2 Expected Results

The performance of the proposed scheduling and link adaptation algorithm was examined using the GPRS software simulator. This simulator is the same with the simulator for the estimation of the performance of the MED algorithm, and the same traffic sources and initial configuration are used. This section presents the results of the simulation for the MAT algorithm.

The results showed that when link utilization increased, the utilization index decreased when the flow was inactive. Therefore, after a long inactivity period, when a new packet arrives at the flow it is served immediately. In summary, for the simulated environment, it could be concluded that the main system limitation was the maximum number of simultaneous TFIs available rather than the packet delays, even under heavy load traffic conditions happened with the VC technique.

In addition, the throughput value for the SMS application was maintained almost constant although the load increased. For this type of application, there is only one packet per session (maximum 8 RB), and as a result the transmission started and finished very quickly once the packet arrived at the transmission queue.

For the MAT strategy, the evolution of the throughput versus the total number of sessions/hour was observed. In this observation the throughput was defined for each user packet as the number of bits divided by the transmission time (queuing time + air interface transmission). At the end of simulation the mean throughput was calculated for all application packets. Besides, the possible retransmissions at the RLC/MAC level were not taken into consideration. From the results, it can be concluded that for the MAT algorithm, the throughput significantly decreased when the traffic increased, falling to values much lower than 10 Kbps for the Web service.

During the simulation the behavior of the MAT strategy was also estimated in terms of the mean waiting time to transfer the first radio block, assuming QoS_1. We noticed that the delay for the MAT strategy was acceptably low. This was due to the evolution of the utilization index parameter in the MAT strategy. For every inactive flow, the MAT algorithm decreased its flow index (equivalent to the timestamp). Then when a new packet of a given flow arrived to the PCU, the first radio block of this flow was immediately served, thus minimizing the delay. To obtain a complete characterization of the performance of the proposed strategy the 95% delay, as defined by ETSI, was also analyzed in this simulation.

Finally, for each pattern of traffic and for the MAT strategy, the maximum number of the sessions/hour admissible meeting QoS delay is reported in Table 3.10.

3.3.9 First-Come First-Served (FCFS)

Another scheduling strategy named FCFS is introduced in this section. An analytic presentation of the technique for GPRS can be found in [21]. The performance of the proposed scheduling technique has also been evaluated in terms of throughput and delay. The simulation results showed that the FCFS as a resource management technique is more suitable for small TBFs than for large flows because it is more probable for a small TBF to be transmitted successfully without any retransmission. For that reason the existence of small TBFs increases the efficiency of the proposed scheduling strategy. On the other hand, the FCFS technique is not suitable for large TBFs, because the number of retransmitted blocks increases rapidly when they have to wait until the end of the transmission of a larger set of blocks, originating from other users. Thus, the efficiency of the FCFS algorithm decreases with large flows.

The efficiency of the proposed algorithm can be increased if a priority factor is inserted into the FCFS main concept. The packet scheduling technique FCFS with priority is more efficient that the pure FCFS, because it gives a transmission priority to the set of blocks that depends only on the number of block retransmissions. The advanced version of the FCFS algorithm is also very simple and easy to be implemented [18].

3.3.9.1 Parameters

The FCFS RRM technique uses the same parameters with the MED technique as presented in Section 3.3.5.2 plus two additional ones. These parameters are the flow time arrival and the priority of the arrived flow. This priority is the transmission priority and depends only on the number of block retransmissions. These two parameters are very important for the FCFS algorithm. The proposed algorithm transmits

Table 3.10   Sessions/Hour for MAT Strategy

Discipline	MAT
Traffic	λ (CS1)
WWW	< 130
E-mail	< 175
SMS	< 90

first the flow that has arrived first or the flow that has the biggest number of block retransmissions. It is obvious therefore that these two parameters have a crucial role in the proper functionality of the algorithm.

3.3.9.2   Expected Results

In this section the results of some performance tests are presented. To estimate the performance of the proposed packet-scheduling algorithm, some evaluating tests took place. In the first evaluation test, the packets were assumed to have a high variance packet size distribution, while in the second one all the packets had the same fixed size.

The simulation environment included a regular cellular layout consisting of nine equally sized three-sector macrosites. Other parameters used for the simulation environment are shown in Table 3.11.

At each time step the C/Is for all active links is calculated. The link level simulations assumed a 16-QAM modulation and ideal frequency hopping. In our simulation the link adaptation was based on the C/I ratio. The choice of modulation and coding schemes was updated every 20 RLC blocks (400 ms).

No admission control algorithm was used in the simulation, which means that for all users that generated packets, there were resources allocated. First, cell selection was performed based on least path loss. Then the user was allocated the channel in the preferred base station with the shortest queue in terms of bits. A user could only had one channel allocated at a time. The WWW traffic model used in the simulation had the parameters presented in Table 3.12.

From the results, it came up that for variable packet size the performance of the FCFS packet scheduling algorithm is not satisfactory. This was expected, since the normalized delay is very dependent on the packet size, which is not taken into account by the FCFS algorithm.

However the performance of the FCFS algorithm increased rapidly in the case of fixed packet size. Especially, for small size packets the performance took its maximum value. On the other hand the performance of the algorithm decreases when the packet size increases. These simulation results confirmed our theoretical assumptions about the FCFS algorithm, that the FCFS favors the small flows, and its performance increased.

3.3.10   Round Robin

In this paragraph we present the round robin scheduling strategy. The performance of the proposed scheduling technique is also evaluated in terms of throughput and delay. The simulation results showed that the round robin as a resource management technique is more suitable for large TBFs than for small flows. Thus the existence of large TBFs increases the efficiency of the proposed scheduling strategy.

Table 3.11   Simulation Environment Parameters

Parameter	Value
Reuse pattern (Regular)	1/3
Distance attenuation	$L = C + 35 \cdot \log(d)$
Log-normal fading, standard deviation	6 dB

Table 3.12 Traffic Model Parameters

Traffic Model Parameters	Value
User interarrival time (Poisson distribution)	Varied
Mean number of packets per session (Geometric distribution)	10 packets
Mean packet interarrival time (Pareto distribution)	10s
Pareto shape parameter	1.4
Mean packet size (log-normal distribution/fixed)	4.1 Kbyte
Standard deviation of packet size	30 Kbyte

Moreover, compared to the FCFS algorithm, the round robin algorithm is more protective and flexible but is much more complicated to implement.

The efficiency of the proposed algorithm can be increased if a priority factor is inserted into the round robin main concept. The packet scheduling technique round robin with priority is more efficient than the pure round robin because it gives a transmission priority to the set of blocks that depends only on the number of block retransmissions. Unfortunately, the advanced version of the round robin algorithm is even more complex and complicated to implement [18].

3.3.10.1 Parameters

The FCFS RRM technique uses the same parameters with the MED technique as presented in Section 3.3.5.2 plus the number of the flows and the number of the time slots. These extra parameters are presented below.

If there are n flows in the ready queue and the time slots are q, then each flow ideally would get $1/n$ of the PDCH time in chunks of q time units, and each flow would wait no longer than nq time units until its next quantum. A more realistic formula would be $n(q + o)$, where o is the context switch overhead. Thus, for practical purposes, it is desirable that the context switch be negligible compared to the time slot.

The performance of the round robin algorithm depends heavily on the size of the quantum (in our case on the size of the packets) used. If the quantum is very large, the round robin algorithm is similar to the FCFS algorithm. If the quantum is very small, the round robin approach is called *processor sharing*.

Another important parameter in the packet scheduling is the mean completion time, which is calculated as follows:

$$\text{Mean Completion Time} = \frac{\sum \text{Time for jobs to complete}}{\text{Number of completed jobs}}$$

Another important parameter of the algorithm is the priority of the arrived flow. This priority is the transmission priority and depends only on the number of block retransmissions. This parameter is very important for the round robin algorithm because the proposed algorithm puts the flow that has the biggest number of block retransmissions in the head of the queue to be served first.

3.3.10.2 Expected Results

The performance of the proposed packet-scheduling algorithm is assessed by performing specific evaluation tests. In the first evaluation test, the packets are assumed to have a high variance for packet size distribution, while in the second one all the packets had the same fixed size.

The simulation environment included a regular cell layout consisting of nine equally sized three-sector macrosites as in the FCFS technique, and all the other parameters including the WWW traffic source model were also the same.

The results show that the throughput performance of the round robin packet-scheduling algorithm increased rapidly in the case of a small number of mobile users. Moreover, in that case the proposed algorithm achieved the best performance in comparison with the other algorithms. However, in the case of a larger number of users the performance of the round robin algorithm was decreased, because in that case the complexity of the algorithm increased the delay.

3.3.11 Least Bits Left First Served

This section introduces the least bits left first served (LBFS) scheduling technique. The concept of this algorithm is based on the fact that longer packets (flows) can tolerate longer delays, and for that reason it is obvious that small flows should be transmitted first. Thus this algorithm favors the small TBFs, because a small TBF is located on the head in the waiting queue and is transmitted first. For that reason the small TBFs increase the efficiency of the proposed scheduling strategy. In addition, the LBFS technique is not suitable for large TBFs, because these blocks are located on the tail of the waiting queue and they have to wait until the end of the transmission of the smaller blocks, originating from other users. Thus, the efficiency of the LBFS algorithm decreases with large flows. The LBFS scheduling technique always transmits the packet with the least number of remaining bits [22]. The performance of the proposed scheduling technique has also been evaluated in terms of throughput and delay.

3.3.11.1 Parameters

The FCFS RRM technique uses the same parameters with the MED technique as presented in Section 3.3.5.2 plus an additional one. This parameter is the number of the bits of the incoming flow. The algorithm calculates the length of any flow (the number of the bits) and inserts the flows into a queue, according to this number. On the head of this queue there exists the flow with the least bits, while in the tail there is the flow that has the biggest number of bits. The proposed algorithm transmits first the packet with the least number of remaining bits. This parameter is very important for the LBFS algorithm.

3.3.11.2 Expected Results

The performance of this packet scheduling algorithm was assessed using the same simulation environment as in the FCFS technique. The simulation environment included a regular cell layout consisting of nine equally sized three-sector macrosites as in the FCFS technique and all the other parameters including the WWW traffic source model were also the same.

From the results it appears that for variable packet size the performance of the LBFS packet-scheduling algorithm increases rapidly. Moreover, in the case of variable packet size the LBFS algorithm achieves the best performance in comparison with the other algorithms. In the case of fixed packet size the performance of the LBFS algorithm is also high, but not as high as in the case of the variable packet size.

3.3.12   Least Time Left First Served

In this section we will present another technique called least time left first served (LTFS). The main concept of this algorithm is that packets (flows) that are close to reaching their QoS criteria should be transmitted first, and for that reason these flows have the maximum priority. On the other hand the flows that are far away from fulfilling their QoS criteria have the least priority. From this short description it is obvious that the LTFS scheduling strategy favors the flows that require high QoS. The performance of the proposed scheduling technique has also been evaluated in terms of throughput and delay.

3.3.12.1   Parameters

The LTFS RRM technique uses the same parameters with the MED technique as presented in Section 3.3.5.2 plus one additional one. This is the normalized delay of the user's last packet, which is used to estimate the time that is left in order to meet the QoS requirements, the LTFS algorithm considers the normalized delay of the user's last packet. After that the algorithm compares the estimated time with the flow's QoS requirements and sorts the flow into the waiting queue according to the comparison results. If the incoming flow originated from a new user then, the LTFS technique considers as estimation parameters the signal strength and the average interference into the system. After that the algorithm compares the estimated time with the flow's QoS requirements and sorts the flow into the waiting queue according to the comparison results.

3.3.12.2   Expected Results

The performance of this packet scheduling algorithm was assessed using the same simulation environment as in the FCFS technique. The simulation environment included a regular cell layout consisting of nine equally sized three-sector macrosites as in the FCFS technique, and all the other parameters including the WWW traffic source model were also the same.

From the results it appears that for a fixed packet size the performance of the LTFS packet scheduling algorithm increases quickly. Moreover, in the case of the fixed packet size the LTFS algorithm achieves the best performance in comparison with other algorithms. In the case of variable packet size the performance of the LTFS algorithm was also acceptable, but not as good as in the fixed packet size case.

References

[1]   3GPP, TSG Services, and System Aspects, "General Packet Radio Service (GPRS): Service Description, Stage 1," 3GPP TS 22.060 (V4.4.0), June 2002.

[2]   http://www.etsi.org.

[3]   http://www.3gpp.org.

[4]   3GPP, TSG Services and System Aspects, "Network Architecture," 3GPP TS 23.002 (V4.7.0), March 2003.

[5]   3GPP, TSG GSM/EDGE, "Mobile Station (MS)—Base Station System (BSS) Interface: Radio Link Control/Medium Access Control (RLC/MAC) Protocol," 3GPP TS 44.060 (V4.11.0), April 2003

[6]   3GPP, TSG GERAN, "Digital Cellular Telecommunications System (Phase 2+): General Packet Radio Service (GPRS); Overall Description of the GPRS Radio Interface; Stage 2," 3GPP TS 43.064 (V4.4.0), April 2003.

[7]   3GPP, TSG Services and System Aspects, "High-Speed Circuit-Switched Data (HSCSD): Stage 1," 3GPP TS 22.034 (V4.1.0), June 2001.

[8]   3GPP, TSG Services and System Aspects, "High-Speed Circuit-Switched Data (HSCSD): Stage 2," 3GPP TS 23.034 (V4.0.0), March 2001.

[9]   3GPP, TSG GSM/EDGE, "Radio Access Network: Multiplexing and Multiple Access on the Radio Path," 3GPP TS 45.002 (V4.6.0), February 2003.

[10] 3GPP, TSG GSM/EDGE, "Radio Access Network: Enhanced Data Rates for GSM Evolution (EDGE): Project Scheduling and Open Issues for EDGE," 3GPP TS 50.059 (V4.0.1), August 2001.

[11] CAUTION++ EU Project, IST–2001–38229, Deliverable D–2.2, "System Requirements Specification," May 2003.

[12] Hong, D., and T. Suda, "Congestion Control and Prevention in ATM Network," *IEEE Network Magazine,* July 1991, pp. 10–16.

[13] Hiroshi Saito, "Optimal Queuing Discipline for Real-Time Traffic at ATM Switching Nodes," *IEEE Transactions on Communications,* December 1990, Vol. 38, No. 12, pp. 2131– 2136.

[14] Ala–Luukko, S., "Mobility Management in IETF and GPRS Specifications," *IEEE Communictions Magazine,* November 1992, pp. 431–435.

[15] Bada, J., and F. Casadevall, "Service Disciplines Performance for GPRS with Link Adaptation and Heterogeneous Traffic," PIMRC 2002. *13th IEEE International Symposium on Personal, Indoor, and Mobile Radio Communications,* Lisboa, Portugal, September 16–18, 2002.

[16] Hyman, J. M., A. A. Lazar, and G. Pacifici, "Real–Time Scheduling with Quality of Service Constraints," *IEEE Journal on Selected Areas in Communications,* Vol. 9, No. 7, September 1991, pp. 1052–1063.

[17] Zhang, L., "Virtual Clock: A New Traffic Control Algorithm for Packet Switching Networks," *ACM Transactions on Computer Systems,* Vol. 9, May 1991, pp. 101–124.

[18] Caffery, J. J., and Stuber, G. L, "Overview of Radiolocation in CDMA Cellular Systems," *IEEE Communications Magazine,* Vol. 36, No. 4, April 1998, pp. 38–45.

[19] Zhang, H., "Service Disciplines for Guaranteed Performance Service in Packet–Switching Networks," *Proc. IEEE,* Vol. 83, No. 10, Oct. 1995, pp. 1374–1396.

[20] Bada, J., et al., "Radio Resource Management in GPRS with Quality of Service," *IST Mobile & Wireless Telecommunications Summit,* Thessaloniki, Greece, June 16–19, 2002.

[21] Kouvatsos, D. D., K. Al–Begain, and I. Awan, "A Queueing Model for a Wireless GSM/GPRS Cell with Multiple Service Classes," *Networking 2002,* Pisa, Italy, May 19–24, 2002.

[22] Waltemar, M., et al., "System–Level Performance Evaluation of Different Scheduling Strategies in EGPRS," *WPMC'01,* Aalborg, Denmark, September 9–12, 2001.

CHAPTER 4
Resource Management in 3G Wireless Systems

Chapters 1–3 presented 2G and 2+G cellular networks as well as a number of radio resource management techniques. Despite the fact that 3G has only recently been launched by a few operators, this chapter gives a good overview of 3G, emphasizing its differences from the previous generations, and describes both the dimensioning steps and the strongest candidate techniques for radio resource management.

4.1 Architecture of 3G Systems

4.1.1 Introduction

3G is a generic term covering a range of current and future wireless network technologies including UMTS, code division multiple access (CDMA) 2000, GPRS, and EDGE. These systems are also known collectively as International Mobile Telecommunications 2000 (IMT–2000), which is a set of proposals defined by the International Telecommunications Union (ITU). In the acronym IMT-2000, the figure 2000 represents both the scheduled year for initial trial systems and the frequency range of 2,000 MHz as it has been defined by the World Administrative Radio Conference that took place in 1992 (WARC-92) in Malaga, Spain.

The most important IMT–2000 standard proposals are UMTS (W-CDMA) as the successor to GSM and CDMA2000 as the interim standard 1995 (IS–95) successor. Both these systems use CDMA technology. This chapter focuses on UMTS, which is considered the most supported IMT-2000 standard; its deployment is already under way in many countries. UMTS will support high-speed packet-switched data (up to 2 Mbps), and it will establish a global roaming standard. In fact, UMTS is the next step beyond the GPRS and is considered to be a transition step from 2G to 3G cellular systems. The UMTS system is being standardized by the 3GPP, which consists of several standard bodies from many countries under the auspices of ETSI. Through UMTS the growing demands for mobile Internet applications that are already an integral part of modern business life are expected to be addressed efficiently.

To reach global acceptance, the 3GPP group is introducing UMTS in phases and annual releases. The first release, which was introduced in December of 1999 (Release '99), defined enhancements and transitions for existing GSM networks. The second phase in 2000 (Release '00) proposed enhancements for IS–95 (with CDMA2000) and TDMA (with TD–CDMA and EDGE). The most important change in Release '99 was the new UTRA, a W–CDMA radio interface for land-based communications. UTRA supports both TDD, which corresponds to a hybrid TDMA/CDMA multiple access scheme, and FDD, which corresponds to the direct sequence CDMA (DS-CDMA) multiple access scheme. The TDD mode is used for public microcells and picocells and unlicensed cordless applications. This is because the TDD mode does not allow large propagation delays between a mobile station and a base station as it would cause a collision between transmit and receive time slots. On the other hand, the TDD mode has the advantage that a large asymmetry of data transfer between uplink and downlink is possible, which is the case for many Internet applications. The FDD mode is used for wide area coverage (i.e., public macro- and microcells. Both modes offer flexible and dynamic data rates up to 2 Mbps for packet-switched connections. The two UMTS modes, used in parallel support the user's requirements efficiently in overlapping application scenarios. The particular features of the FDD are listed as follows:

- Microcell and macrocell coverage;
- Optimal symmetric access;
- Up to 364-Kbps capacity;
- Supportive of high mobility;
- Best fit for wide area coverage.

The TDD mode features are listed as follows:

- Picocell and microcell coverage;
- Optimal symmetric and asymmetric access;
- A 2-Mbps capacity;
- Supportive of low mobility;
- Best fit for high traffic requirements.

We have to mention also that the FDD technique requires paired spectrum whereas the TDD technique can use unpaired spectrum. Paired spectrum means that a communication takes place using a part of the spectrum in a lower frequency band, and another part in an upper frequency band. Paired spectrum is usually specified in a form like "2×15 MHz" meaning 15 MHz in the lower band and 15 MHz in the upper band. The lower band is used for uplink and the other one for downlink transfers.

UMTS allows many more applications, in comparison with GSM, to be introduced to a worldwide base of users and addresses the growing demand of mobile Internet applications for new capacity. This is the main advantage of UMTS from the point of view of the user, who will experience a broad offer of user-friendly and speedy services. In fact all the services that are now available for normal Internet users, like WWW browsing and multimedia communications in general will be also available for users while on the move.

The UMTS services are divided into three abstract layers. There is the applications layer, the services layer, and the network services layer. The applications layer, which is not a standardized part of UMTS, is what the end user actually uses (for such functions as e-mail and videoconferencing.) The services layer is what offers itself to the applications layer. The services layer is standardized and is comprised of various components. This means that new services can be performed with a combination of old components in a way similar to IN-services. The three main components are connection-oriented communication, connectionless communication, and applications management.

To increase the variety of offered services and the competition between operators, 3GPP defines only the so-called bearer services by standardizing for each of them network parameters like bit rate, bit error rate, and delay. From the users' point of view the application that is running on their terminals is called a teleservice, and it can make use of several bearer services.

The framework for service description in UMTS is comprised of three categories: (1) the basic telecommunication services, which are further distinguished into bearer services and teleservices; (2) the supplementary services; and (3) the service features.

The bearer services provide the capability for information transfer between access points and involve only low-layer functions. They also define the information transfer attributes like connection mode and bit rate and the quality of the connection attributes like delay and bit error rate (BER). UMTS teleservices and supplementary services are built independently by each service provider or network operator by exploiting the capabilities of the network and the user terminals. However there exist four UMTS teleservices that are being standardized completely by ETSI; these are speech, fax, SMS, and emergency calls. Provision for GSM phase 2+ supplementary services like call forwarding, call waiting, and call hold should also take place in UMTS in a way that is transparent to the user. Finally the UMTS service features define specific building blocks that can be used to create services. Such building blocks are incorporated as functionalities and mechanisms in architectures such as customized application for mobile network enhanced logic (CAMEL), mobile execution environment (MexE), and intelligent network (IN). CAMEL enables worldwide access to operator-specific IN applications such as prepaid mobile services, call screening, and active network supervision. CAMEL is the primary GSM Phase 2+ enhancement for the introduction of the UMTS virtual home environment (VHE) concept.

It is possible to negotiate and renegotiate the characteristics of a bearer service during a session or during the connection establishment of a session. Both connection-oriented and connectionless services are offered for point-to-point and point-to-multipoint communication. Bearer services have different QoS parameters for optimizing transfer delay, delay variation, and bit error rate. The offered data rate targets are the following:

- 144-Kbps satellite and rural outdoor;
- 384-Kbps urban outdoor;
- 2,048-Kbps indoor and low range outdoor.

The UMTS standards define different QoS classes for four types of traffic as follows:

- Conversational class (voice and video telephony);
- Streaming class (multimedia, video on demand, and Web casts);
- Interactive class (Web browsing, network gaming, and database access);
- Background class (e-mail, SMS, and downloading).

UMTS offers also a capability known as VHE. It is a concept for personal service environment portability across network boundaries and between terminals. It provides to users the ability to tailor their services, in a seamless manner, across networks and operators, independently of the different supporting environments. VHE will enable terminals to negotiate functionality with the visited network, possibly by a software download, and "home-like" services will be provided. In that way users are consistently presented with the same personalized features, user interface customization, and services at any given time. Within this framework UMTS provides also improved network security and location-based services.

The core network of UMTS is in fact an evolution of the GSM core network. The critical innovation is the radio access network, especially the method of radio transmission, which is completely new since it is based on CDMA. However the introduction of UMTS will maintain backward compatibility with GSM, which will not only continue to be used but will also evolve. Thus there will exist a common core network with two independent radio access networks: one for UMTS and one for GSM. The UMTS radio access network will allow for multimedia applications because of the larger bandwidth of the radio channels (5 MHz instead of 200 kHz in GSM) and the new access method.

4.1.2   UMTS Cell Structure and Spectrum Usage

UMTS was conceived from the beginning to offer global radio coverage and worldwide roaming. It is envisaged as a global system, comprising both national terrestrial and global satellite components. It uses a hierarchical cell structure comprised of micro-, macro-, and picocells as depicted in Figure 4.1. The operators use such

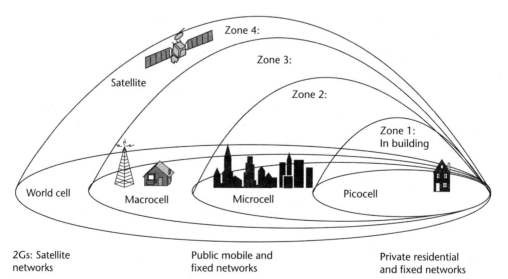

Figure 4.1   UMTS coverage.

hierarchical structures to optimize their networks' deployment in terms of both capacity and coverage.

As shown in previous chapters, smaller cells allow for a higher user density. Therefore macrocells are usually deployed for land-wide coverage; microcells are installed in areas with higher population density (as it is the case with normal 2G GSM cells), and picocells are used in buildings and for hot spot areas like stadiums, airports, and railway stations. In that way the requirement to achieve truly personal communication, with terminals able to roam from a WLAN or even a fixed network, into a pico- or microcellular public network, then into a wide area macrocellular network, and then to a satellite mobile network is accomplished. Of course the most important aspect for the user during this roaming procedure is to carry out a seamless communication session when moving from the one network to the other.

An important aspect that should be carefully treated by network planners is the type of the transmission mode (FDD or TDD) that should be employed in each cell type. As we have seen the TDD mode is optimized for public micro- and picocells and unlicensed cordless environments. The FDD mode is optimized for wide area coverage (i.e., public macro- and microcells). A hierarchical structure consisting of three layers is shown in Figure 4.2 [1]. It consists of pico-, micro-, and macrocells. The macrocells are used for wide area coverage whereas the microcells are used as in normal GSM to provide extra capacity where macrocells could not cope. The picocells are deployed mainly indoors for supporting high data rate services.

The spectrum for UMTS must be sufficient to handle the traffic requirements to give to operators confidence in the possibilities for developing and introducing 3G

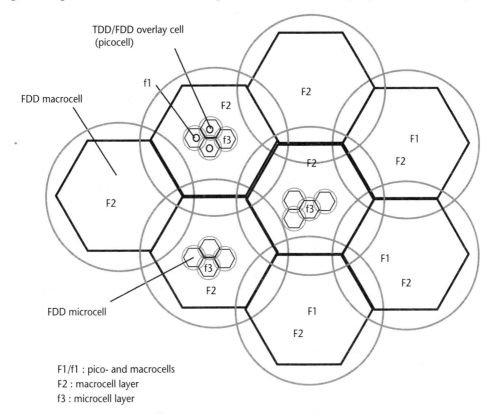

F1/f1 : pico- and macrocells
F2 : macrocell layer
f3 : microcell layer

Figure 4.2   Typical UMTS cell structure.

services. The UMTS Forum recommends 2 × 15 MHz (paired) plus 5-MHz (unpaired) as the preferred minimum spectrum requirement for each public UMTS operator in the initial phase. It is believed that the allocation of unpaired spectrum will handle asymmetric traffic in an optimized way. However, depending on country-specific situations, other spectrum allocations per operator may be more appropriate.

The spectrum available for operators is was identified at WARC-92 and in the subsequent ITU radio regulations. Figure 4.3 shows the ITU frequency allocations for IMT-2000. These allocations have been further enhanced at WRC 2000, which identified the bands 1,710–1,885 and 2,500–2,690 MHz also for IMT-2000.

The paired bands that are used by UTRA/FDD for transmission and reception are the following [2]:

1. 1,920 to 1,980 MHz used for uplink (mobile transmit and base receive) and 2,110 to 2,170 MHz used for downlink (base transmit and mobile receive);
2. 1,850 to 1,910 MHz used for uplink (mobile transmit and base receive) and 1,930 to 1,990 MHz used for downlink (base transmit and mobile receive).

Usually the unpaired spectrum is configured into 5-MHz packages and the paired spectrum in 2 × 5 MHz and 2 × 10 MHz packages. Paired spectrum of 2 × 15 is also allocated when there are predictions for high traffic demands in a particular area. This allocation allows for a complete hierarchical cell structure to be deployed. On the other hand, 2 × 5 MHz will allow a single layer only, and thus a hierarchical cell structure is not feasible in this case. An allocation of 2 × 10 MHz gives room for a two-layer structure (e.g., a macrocell layer together with either a microcell layer or a picocell layer). In addition to the allocation of paired frequencies an operator usually needs the allocation of unpaired frequencies for TDD operation. This is required mainly for satisfying low mobility applications running indoors and is often configured into 5-MHz packages as mentioned above to give satisfactory capacity for asymmetric traffic [1].

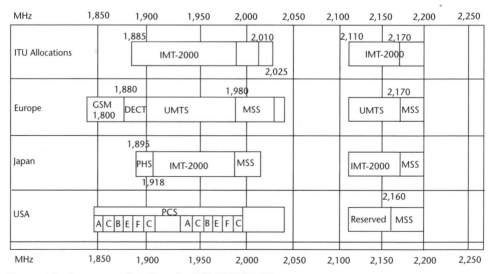

Figure 4.3 Frequency allocations for IMT-2000/UMTS.

4.1.3   UMTS Network Architecture

A UMTS network consists of two interacting domains: the infrastructure domain consisting of the core network (CN) and the UTRA network (UTRAN) and the user equipment (UE) domain. The UTRAN is comprised of the mobile station, the base station (transceiver, antenna, and controller), and the radio interface between them. As we have mentioned before the UMTS core network is an evolution of the GSM one. In particular UMTS (Release '99) incorporates the enhanced GSM Phase 2+ core network with GPRS and CAMEL. This enables network operators to enjoy the improved cost efficiency of UMTS while protecting their 2G investments and reducing the risks of implementation. Of course all equipment has to be modified for UMTS operation and services. The main function of the core network is to provide switching and routing for user traffic. The core network also contains the required databases and network management functions. The core network domain is subdivided into the serving network domain, the home network domain, and the transit network domain. Figure 4.4 shows a simplified UMTS architecture with its basic domains and external reference points and interfaces with the UTRAN.

The UTRAN, is connected via the Iu interface to the GSM Phase 2+ CN. The Iu is the UTRAN interface between the radio network controller (RNC) and the CN. The UTRAN interface between RNC and the packet-switched domain of the CN is called Iu–PS, and the UTRAN interface between RNC and the circuit-switched domain of the CN is called Iu–CS. User equipment has a radio interface to the UTRAN that is called Uu. Note that many times the previously mentioned interfaces are also called reference points. The UE domain will be further analyzed in Section 4.1.3.1 and the UTRAN architecture in Section 4.1.3.2.

4.1.3.1   UMTS User Equipment

The user equipment domain may appear with a variety of equipment types with different levels of functionality. They may be compatible with one or more existing access (fixed or radio) interfaces and may include a removable smart card to be used in different user equipment types. The user equipment is further subdivided into the mobile equipment domain (ME) and the user services identity module domain (USIM) as shown in Figure 4.5. The reference point between the ME and the USIM

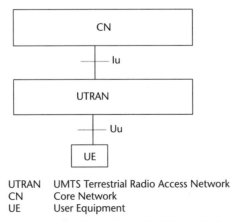

UTRAN	UMTS Terrestrial Radio Access Network
CN	Core Network
UE	User Equipment

Figure 4.4 UMTS architecture. (*Source:* ETSI. Reprinted with permission.)

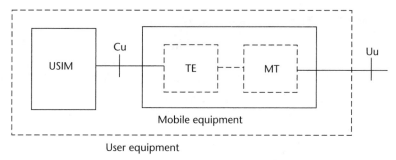

User equipment

Figure 4.5   User equipment.

is called Cu. The mobile equipment may be further subdivided into several entities. The typical entities in the ME domain are the mobile termination (MT), which performs radio transmission and related functions, and the terminal equipment (TE), which contains the end-to-end applications [3].

The UMTS UE is based on the same principles as the GSM MS. The mobile equipment performs radio transmission and contains applications. The USIM contains data and procedures that unambiguously and securely identify it. These functions are typically embedded in a standalone smart card. This device is associated with a given user, and as such it can identify this user regardless of the ME he or she uses.

4.1.3.2   UTRAN

As mentioned above, UMTS differs from GSM Phase 2+ mostly in the new principles for air interface transmission, which are incorporated in the UTRAN. The UTRAN consists of a set of radio network subsystems (RNSs), which are the access parts of the UMTS network. A RNS offers the allocation and the release of specific radio resources to establish a connection between an UE and the UTRAN. An RNS is connected to the core network through the reference interconnection point Iu, and it consists of two new network elements, namely the RNC, and node B. The RNC is connected to a set of node B elements, each of which can serve one or several cells [4].

The RNC has the overall control of the logical resources of the node Bs that belong to it, and it is also responsible for the handover decisions that require signaling to the UE. A node B is connected to the RNC through the Iub interface. Inside the UTRAN, the RNCs of the RNSs are interconnected through the Iur interface. The Iur interface is implemented either through a physical direct connection between RNCs or via a suitable transport network. Each RNS is usually responsible for the resource management and transmission and reception in more than one cell.

Network elements from GSM phase 2+, such as MSC, SGSN, and HLR, are extended to adopt the UMTS requirements, but the RNC, node B, and the mobile terminals are designed from scratch. RNC is the replacement for BSC, and node B corresponds to the BTS in terms of functionalities. GSM and GPRS networks are extended, and new services are integrated into an overall network that contains both existing interfaces such as A, Gb, and A_{bis}, and new interfaces. The overall architecture of the new system is presented in Figure 4.6.

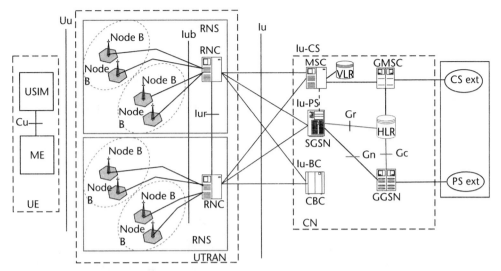

Figure 4.6   UTRAN and overall UMTS system architecture.

UMTS defines four new interfaces or reference points:

- Uu: UE to node B (UTRA, the UMTS W–CDMA air interface;
- Iu: RNC to GSM Phase 2+ CN interface (MSC/VLR or SGSN);
  - Iu-CS for circuit-switched data;
  - Iu-PS for packet-switched data;
- Iub: RNC to node B interface;
- Iur: RNC to RNC interface, not comparable to any interface in GSM.

The Iu, Iub, and Iur interfaces are based on ATM transmission principles. The RNC enables autonomous RRM. It performs the same functions as the GSM BSC, providing central control for the RNS elements (RNC and node Bs). The RNC handles protocol exchanges between Iu, Iur, and Iub interfaces and is responsible for centralized operation and maintenance (O&M) of the entire RNS with access to the OSS. Because the interfaces are ATM-based, the RNC switches ATM cells between them. The user's circuit-switched and packet-switched data coming from Iu–CS and Iu–PS interfaces are multiplexed together for multimedia transmission via Iur, Iub, and Uu interfaces to and from the UE.

As we mentioned above the RNS uses the Iur interface for RRM purposes. For each connection between the UE and the UTRAN, one RNS is the serving RNS. Serving control functions such as admission control, congestion control, and handover are managed entirely by a single serving RNS. When a UE must use resources in a cell not controlled by its serving RNS, the serving RNS must ask the controlling RNS for those resources. This request is made via the Iur interface, which connects the RNSs with each other. In this case, the controlling RNS is also said to be a drift RNS for this particular UE as shown in Figure 4.7. This kind of operation is needed for providing soft handover throughout the network.

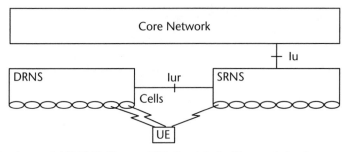

Figure 4.7   Serving and drift RNS. (*Source:* ETSI. Reprinted with permission.)

Soft handover is one of the most important aspects of UMTS. For user terminals in the soft handover process, the original serving base station and the target base station will maintain two communications links simultaneously over the same bandwidth to guarantee a smooth transition without the chance of dropping the ongoing call. The condition of having more than one radio link active at the same time is called macro diversity. This flexibility in keeping the connection open to more than one base station results in fewer lost calls, which is very important for the operator.

A node B is a logical node responsible for radio transmission/reception in one or more cells to/from the UE, in other words it can be considered as a network component that serves one cell as it happens with the BTS in GSM. It can support FDD mode, TDD mode, or dual-mode operation, and there is only one RNC for any node B. node B connects with the UE via the W–CDMA Uu radio interface and with the RNC via the Iub ATM–based interface. Node B is the ATM termination point, and it can be colocated with a GSM BTS to reduce implementation costs.

Node B's functions include radio and modulation/spreading aspects along with channel coding (forward error correction) and some combining/splitting functions for soft handover. It is also converting the data flows between the Iu-b and the Uu interfaces and participates in radio resource management. There are two chip-rate options when node B is operating in the TDD mode: 3.84 Mcps TDD and 1.28 Mcps TDD and each TDD cell support either of these options. A node B that supports TDD cells can support one chip-rate option only, or both options. A RNC that supports TDD cells can support one chip-rate option only, or both options. The nominal channel spacing for 3.84 Mcps TDD is 5 MHz and for 1.28 Mcps TDD is 1.6 MHz. However the distance between channels can be adjusted accordingly to optimize performance in a particular deployment scenario [5].

4.1.3.3   UMTS Interfaces

As discussed in Section 4.1.3.2 UMTS defines four new interfaces: Uu, Iub, Iur, and Iu. These interfaces owe their existence to the new air interface and may be referred to as either UMTS interfaces or UTRAN interfaces. The general protocol model for UTRAN interfaces is shown in Figure 4.8. It consists of a set of horizontal and vertical layers. The structure is based on the principle that the layers and planes are logically independent of each other. Therefore, it is easier for the standardization bodies to alter the protocol stacks in order to fit future requirements [4].

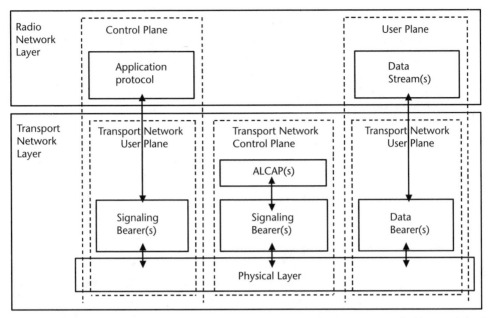

Figure 4.8   General protocol model for UTRAN interfaces. (*Source:* ETSI. Reprinted with permission.)

The general protocol model consists of two main layers, the radio network layer, and transport network layer. All UTRAN-related requirements are addressed only in the radio network layer, and the transport network layer represents standard transport technology that is selected for usage in UTRAN, but without any UTRAN specific requirements. In the vertical direction we have a set of control and user planes. Control planes are used to control a link or a connection; user planes are used to transparently transmit user data from the higher layers.

The control plane (CP) consists mainly of the application protocol and the signaling bearers. The application protocol is used for setting up bearers for (i.e., radio access bearer or radio link) in the radio network layer. The signaling bearer is used for transporting the application protocol messages. It may or may not be of the same type as the signaling protocol for the access link control application part (ALCAP) and is always set up by O&M actions. ALCAP is a generic name for the transport signaling protocol that is reacting to the radio network layer's demands to set up, maintain, and release data bearers.

The user plane (UP) consists of the data streams and the data bearers for the data streams. Data streams contain the user data that is transparently transmitted between the network elements. Data bearers are the frame protocols used to transport user data.

The transport network control plane (TNCP) does not include any radio network layer information and is completely in the transport layer. It includes the ALCAP protocols that are needed to set up the data bearers for the user plane. It also includes the appropriate signaling bearers needed for the ALCAP protocols. This plane acts between the control plane and the user plane and allows the application protocol in the radio network control plane to be completely independent of the technology selected for the data bearer in the user plane.

It should be noted that ALCAP might not be used for all types data bearers. The TNCP is present in the Iu–CS, Iur, and Iub interfaces. In the remaining interfaces where there is no ALCAP signaling, preconfigured data bearers are activated. The signaling bearer for the ALCAP may or may not be of the same type as the signaling bearer for the application protocol. The signaling bearer for ALCAP is always set up by O&M actions.

The transport network user plane (TNUP) consists of the data bearers in the user plane, and the signaling bearers of the application protocol in the control plane. The data bearers in TNUP are directly controlled by the TNCP during real-time operation, but the control actions required for setting up the signaling bearers for application protocol are done through O&M actions.

4.1.3.3.1   Iu Interface: General Aspects and Principles

The UMTS Iu interface enables interconnection of radio network controllers (RNCs) with core network nodes. It is an open interface and divides the system so that CN handles switching, routing, and service control, and UTRAN handles radio resource management. The Iu interface toward the PS-domain of the core network is called Iu-PS, and the Iu interface toward the CS-domain is called Iu-CS. The Iu interface to the broadcast domain is called Iu-BC (see Figure 4.6) [6].

There shall not be more than one Iu-PS interface toward the PS-domain from any one RNC and each RNC shall not have more than one Iu-CS interface toward its default CN node within the CS domain. However it may also have further Iu-CS interfaces toward other CN nodes within the CS domain. It should be noted that an RNC has one single permanent default CN node per CN domain.

The Iu interface supports the following procedures and functionalities:

1. The establishment, maintenance and release of radio access bearers;
2. Serving radio network subsystem (SRNS) relocation, intrasystem handover, intersystem handover, and intersystem change;
3. Procedures to support the cell broadcast service;
4. The separation of each UE on the protocol level for user specific signaling management;
5. Location services by transferring requests from the CN to UTRAN, and location information from UTRAN to CN;
6. Simultaneous access to multiple CN domains for a single UE;
7. Mechanisms for resource reservation for packet data streams.

The application protocol used in the Iu interface is called radio access network application part (RANAP) and is responsible for several functions and procedures [7]. The transport protocol used for both Iu-CS and Iu-PS is ATM while TCP/IP is used as the bearer for the radio network layer protocol over Iu-BC. The detailed protocol specifications of the Iu are given in 3GPP specifications 25.411 to 25.419. The Iu interface corresponds to the A interface of GSM.

4.1.3.3.2   Iur Interface: General Aspects and Principles

The Iur interface connects two RNCs in the UTRAN and has no equivalent in GSM system. It uses also ATM as the transport protocol. The basic capabilities of Iur are:

(1) support of inter-RNC mobility, (2) support of dedicated channel traffic between two RNCs, and (3) support of common channel traffic between two RNCs [8].

The main functions on the Iur interface are the following:

1. Transport network management;
2. Traffic management of common transport channels;
3. Traffic management of dedicated transport channels;
4. Traffic management of downlink shared transport channels and TDD-uplink shared transport channels when applicable;
5. Measurement reporting for common and dedicated measurement objects.

The previously listed functions may include several subfunctions. The application protocol used in the Iur interface is called the radio network subsystem application part (RNSAP) and is responsible for providing signaling information across it [9].

RNSAP procedures are divided into four categories as follows:

1. RNSAP basic mobility procedures;
2. RNSAP dedicated transport channel (DCH) procedures;
3. RNSAP common transport channel procedures;
4. RNSAP global procedures.

4.1.3.3.3   Iub Interface: General Aspects and Principles

As we have mentioned before the Iub is the logical interface that connects a node B with an RNC [10].

The main functions on the Iub interface are the following:

1. Management of Iub transport resources;
2. Logical O&M of node B;
3. Implementation-specific O&M transport;
4. System information management;
5. Traffic management of common channels;
6. Traffic management of dedicated channels;
7. Traffic management of shared channels;
8. Timing and synchronization management.

The previously listed functions may include several subfunctions. The Application Protocol used in the Iub interface is called node B application part (NBAP). The specification of signaling transport related to NBAP to be used across the Iub Interface is given in [11].

4.1.3.3.4   The Radio Interface Uu

The UMTS radio interface corresponds to the Uu reference point, which provides interconnection between RNC and the user terminal via the node B. The radio interface is layered into three protocol layers, namely the physical layer (L1), the data link layer (L2), and the network layer (L3) [12]. Figure 4.9 shows the radio interface protocol architecture.

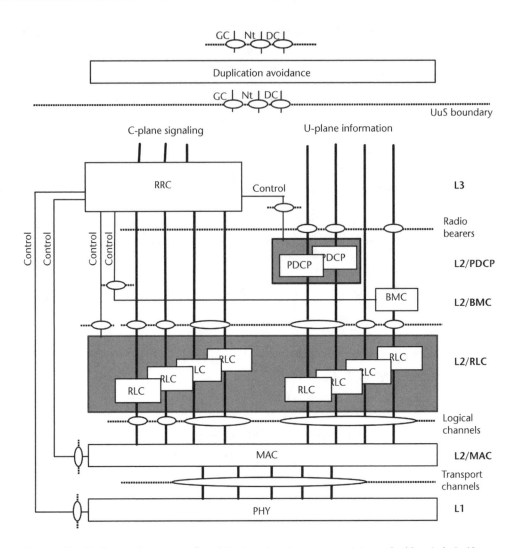

Figure 4.9   Radio interface protocol architecture (service access points marked by circles). (*Source:* ETSI. Reproduced with permission.)

The data link layer is split into the following sublayers: MAC, RLC, packet data convergence protocol (PDCP), and broadcast/multicast control (BMC).

The MAC sublayer is located on top of the physical layer. Logical channels are used for communication with the higher layers, and transport channels are used for exchanging information with the physical layer. A set of logical channels is defined to transmit a specific type of information. Therefore, a logical channel determines the kind of information it uses. The transport channels describe how data is to be transmitted over the air interface and with what characteristics [13].

There exist three types of MAC entities in the MAC sublayer:

1. MAC-b is the MAC entity that handles the following transport channels:

   • *Broadcast channel (BCH)*: A downlink channel used for broadcast of system information into an entire cell.

2. MAC-c/sh is the MAC entity that handles the following transport channels:

- *PCH:* A downlink channel used for broadcast of control information into an entire cell allowing efficient UE sleep mode procedures. Currently identified information types are paging and notification. Another use could be UTRAN notification of change of BCCH information.
- *Forward access channel (FACH):* Common downlink channel without closed-loop power control used for transmission of relatively small amount of data;
- *Random access channel (RACH):* A contention-based uplink channel used for transmission of relatively small amounts of data (e.g., for initial access or nonreal-time dedicated control or traffic data);
- *Common packet channel (UL CPCH):* A contention-based channel used for transmission of bursty data traffic, this fast power–controlled channel only exists in FDD mode and only in the uplink direction (since it is shared by the UEs in a cell and thus a common resource);
- *Downlink shared channel (DSCH):* A downlink channel shared by several UEs carrying dedicated control or traffic data:
- *Uplink shared channel (USCH):* An uplink channel shared by several UEs carrying dedicated control or traffic data, used in TDD mode only.

3. MAC-d is the MAC entity that handles the following transport channels:

- Dedicated transport channels (DCHs): The channels dedicated to one UE used in uplink or downlink.

The union of the channels mentioned in the three different MAC entities above form the set of transport channels in UMTS, which are provided through specific service access points (SAPs) between the MAC and the physical layer.

The RLC sublayer is responsible for such functions as acknowledged or unacknowledged data transfer, establishment of RLC connections, transparent data transfer, QoS settings, and unrecoverable error notification. There is one RLC connection per radio bearer [14].

The PDCP sublayer is responsible for the transmission and reception of radio network layer protocol data units (PDUs). Within UMTS, several different network layer protocols are supported to transparently transmit protocol data. For the moment, IPv4 and IPv6 are supported, but UMTS is open to other protocols without forcing the modification of UTRAN protocols. This transparent transmission is one task of PDCP; another is to increase channel efficiency (e.g., by protocol header compression). It is used only in the user plane [15].

The BMC sublayer offers broadcast/multicast services in the user plane. For instance, it stores SMS CB messages and transmits them to the UE. It is also used only in the user plane [16].

The MAC sublayer provides data transfer services on logical channels. A set of logical channel types is defined for different kinds of data transfer services as offered by MAC, and each logical channel type is defined by what type of information is transferred. The logical channels are provided through specific SAPs between RLC and the MAC sublayer, and they are split into two categories: control and traffic channels.

The control channels, which are used for transfer of control plane information only, are listed as follows.

- BCCH, downlink only;
- PCCH, downlink only;
- CCCH, uplink and downlink;
- Dedicated control channel (DCCH), uplink and downlink;
- Shared channel control channel (SHCCH), uplink and downlink.

The traffic channels, which are used for the transfer of user plane information only, are the following:

- Dedicated traffic channel (DTCH), uplink and downlink;
- Common traffic channel (CTCH), downlink only.

Layer 3 and RLC are divided into control (C-) and user (U-) planes. PDCP and BMC exist in the U-plane only. In the C-plane, layer 3 is partitioned into sublayers where the lowest sublayer, denoted as radio resource control (RRC), interfaces with layer 2 and terminates in the UTRAN.

The next sublayer provides duplication avoidance functionality, which prevents the reception of duplicated messages. The C-plane radio bearers, which are provided by RLC to RRC, are denoted as signaling radio bearers. In the C-plane, the interface between duplication avoidance and higher L3 sublayers is defined by the general control (GC), notification (Nt), and dedicated control (DC) SAPs.

In Figure 4.9 the connections between RRC and MAC as well as RRC and L1 providing local interlayer control services are also shown. An equivalent control interface exists between RRC and the RLC sublayer, between RRC and the PDCP sublayer, and between RRC and BMC sublayer. These interfaces allow the RRC to control the configuration of the lower layers. For this purpose separate control SAPs are defined between RRC and each lower layer (PDCP, RLC, MAC, and L1).

Finally there exist specific mapping rules between the logical channels and the transport channels. In particular, in the uplink direction, the following mappings between logical channels and transport channels are possible:

- CCCH can be mapped to RACH;
- DCCH can be mapped to RACH;
- DCCH can be mapped to CPCH (in FDD mode only);
- DCCH can be mapped to DCH;
- DCCH can be mapped to USCH (in TDD mode only);
- DTCH can be mapped to RACH;
- DTCH can be mapped to CPCH (in FDD mode only);
- DTCH can be mapped to DCH;
- DTCH can be mapped to USCH (in TDD mode only);

- SHCCH can be mapped to RACH (in TDD mode only);
- SHCCH can be mapped to USCH (in TDD mode only).

In the downlink, the following mappings between logical channels and transport channels are possible:

- BCCH can be mapped to BCH;
- BCCH can be mapped to FACH;
- PCCH can be mapped to PCH;
- CCCH can be mapped to FACH;
- DCCH can be mapped to FACH;
- DCCH can be mapped to DSCH;
- DCCH can be mapped to DCH;
- DTCH can be mapped to FACH;
- DTCH can be mapped to DSCH;
- DTCH can be mapped to DCH;
- CTCH can be mapped to FACH;
- SHCCH can be mapped to FACH (in TDD mode only).
- SHCCH can be mapped to DSCH (in TDD mode only).

The previously mentioned channels can be controlled in accordance with the service that is going to be deployed to achieve its requirements. The RRC block plays a vital role toward achieving this objective since it incorporates QoS control functionalities and RRM operations among others. Over the air interface, RRC messages carry all the relevant information required for setting up a signaling radio bearer (during the lifetime of the RRC connection) and setting up, modifying, and releasing radio bearers between UE and UTRAN (all being part of the RRC connection).

4.1.4   Evolution of GSM Toward UMTS

An important step toward UMTS is be the widespread deployment of packet radio services being deployed over 2G systems, such as GPRS for GSM. These systems are giving valuable experience with connectionless systems for the operators and provide a platform for the development of service interworking functions and service provider interfaces as well as a core of mobile multimedia services. What is most important is that this is done with less initial investment than it is necessary for UMTS, where a completely new radio infrastructure and terminals are needed. Customers are attracted to these intermediate networks by the provision of attractive services from new content and service providers. These customers will then be willing to invest in new UMTS terminals in the anticipation of better and more efficient delivery of these services. In turn, this provides the incentive for network operators to invest in UMTS infrastructure to satisfy the need for capacity demanded by a successful mobile multimedia mass market [17].

The steps toward full deployment of UMTS are: (1) incorporation of packet switching capabilities onto the existing GSM networks together with the appropriate attractive services; (2) initial trials of prototype UMTS nodes; (3) basic deployment phase beginning in 2002, which includes the first incorporation of UTRA base stations into operational networks and support of both narrowband and broadband services over the same UTRA interfaces; and (4) full commercial phase, beginning in 2003 and achieving the envisaged performance and capabilities by 2005.

The deployment of UTRANs will start gradually in hot spot areas and in cooperation with GSM. GSM BSS access networks will be a key element for service continuity in UMTS networks. The UMTS phase 1 network is planned as a direct evolution of the GSM core network. Full coverage will be provided in later years. The VHE environment will play a vital role toward this by allowing 3G functionalities in 2G systems. So UMTS phase 1 shall be developed in such a way that it supports compatibility with an evolved GSM network from the point of view of roaming and handover. This could be achieved by evolving from a GSM phase 2+ network but does not exclude other developments. UMTS phase 1 shall in particular support bursty and asymmetric traffic in an efficient way. This shall allow UMTS phase 1 to support single- and multimedia N-ISDN applications and single- and multimedia IP applications [18]. An overall UMTS system approach is needed for UMTS phase 1 development as it is more than the addition of a UTRAN to a GSM phase 2+ architecture. Requirements to the GSM phase 2+ core network for UMTS should be incorporated. Such a hybrid 2G-3G configuration is depicted in Figure 4.10.

As far as future phases of UMTS are concerned some key characteristics have been already defined with the first priority being the achievement of an all-IP architecture. This is already fulfilled in the release 5 of 3G specifications from 3GPP, which is frozen but not yet deployed. Release 5 includes also security and integrity functionalities and end-to-end QoS messaging enhancements. At the same time 3GPP is already working toward the release 6 of 3G specifications. The list of new functionalities envisaged for this release includes: usage of multiple input multiple output (MIMO) antennas, security enhancements, global roaming and global interoperability, advanced terminals that can roam across different heterogeneous networks with power saving capabilities, virtual network operators, and the ability

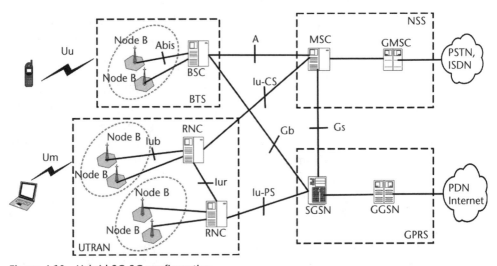

Figure 4.10 Hybrid 2G-3G configuration.

to offer flexible bandwidth on demand and asymmetric bandwidth. All new features and characteristics corresponding to the various 3G releases can be found in [19]. Release 6 specifications are expected to be frozen in late 2003 and deployed after about 3 years.

4.2 UMTS Dimensioning

This section describes the basic planning and dimensioning guidelines that the operator should follow for the successful deployment of a UMTS network. This is quite important for the understanding of the procedures that allow the correct site selection and for the proper configuration of the various parameters. It has to be noted that no radio resource management mechanisms are applied at this stage, while these can be applied at a next step, after the network is set up.

4.2.1 Planning Guidelines

Radio network planning process for a 3G system can be divided into two phases: system dimensioning and detailed radio planning [20]. In a WCDMA cellular system where all connections operate on the same carrier, the number of users that can be served simultaneously, influences the receiver's noise floor. Therefore, in such a system, like UMTS, the planning phases cannot be separated into coverage and capacity planning. Instead of that, the main task is the prediction of the traffic profile, so that adequate resources are provisioned for each area.

In 2+G cellular systems data services started to play an important role, especially after the introduction of GPRS, as described in Chapter 3. Service provision requires the complete optimization process to overcome a set of constraints. An important task is the definition of the QoS requirements for different service categories. Until now, it has been adequate to specify the coverage and blocking probability mainly for voice service, while for GPRS the dimensioning also includes the constant provision of data services.

In UMTS, dimensioning is more complicated, since indoor and in-car coverage probabilities have to be considered too. In addition, the problem can be described as multidimensional; for each service the QoS targets have to be set and naturally also met. Together with the coverage probability and other known measures for real-time traffic, the packet data QoS criteria have to be defined. UMTS penetration and the service distribution over GSM traffic, as well as outdoor/in-door/in-car subscriber distribution are to be considered too, before the planning and simulation takes place. Table 4.1 presents the requirements set for three basic services of UMTS, namely Voice 12.2, PS 64/64, and PS 128/64. These requirements focus on the air interface blocking and the coverage requirements.

Table 4.1 Services and QoS Requirements

Service	UL Bit Rate (Kbps)	DL Bit Rate (Kbps)	Air Interface Blocking	Coverage Requirements
Voice 12.2	12.2	12.2	< 2%	> 95%
PS 64	64	64	< 2%	> 90%
PS 128	64	128	< 2%	> 90%

While in 2G systems planning is strongly concentrated in coverage dimensioning, a more detailed interference planning and capacity analysis than simple coverage optimization is needed in 3G systems. For that purpose radio network planning tools are utilized. These aim to assist the planner to optimize the base station parameters, the site selections and antenna directions, and even the site locations, in order to meet the QoS and the capacity and service requirements at a minimum cost. To achieve the optimum result the tool must have knowledge of the radio resource algorithms to perform the corresponding operations and take the appropriate decisions. Uplink and downlink coverage probability is determined for a specific service by testing the service availability in each location. The mobile station density in different cells should be based on actual traffic information. The hotspots should be identified as an input for accurate analysis. A source of information concerning user density would be the data coming from the operator of the 2G network. Planning should keep GSM sites with respect to the defined hotspots. Therefore, the planner should try to start with the best 3 grid-sector/site. Furthermore, pilot pollution should be minimized, so that the UMTS cell dominance is isolated. The cell traffic has to be balanced by the trade-off between throughput and interference. The soft handover areas are intended to be kept within limits and the coverage and capacity requirements should be highlighted.

The coverage and capacity requirements within urban and dense urban environments, will directly lead to high site densities. Correspondingly, a microcell deployment will be an attractive solution in terms of relative ease of site acquisition, increased air interface capacity, and more efficient indoor penetration.

4.2.2 Dimensioning Example

This example presents the dimensioning of a 3G system based on the existing 2G network [21]. It is assumed that the operator utilizes a GSM network, and a UMTS/FDD and UMTS/TDD infrastructure. Figure 4.11 presents the current GSM network configuration in the area of interest.

The area of interest comprises of a hotspot area of buildings, local shopping centers and amusement complexes. The areas around the eclipse are mainly domestic, where traffic load is not expected to be heavy.

The most important requirement for any radio network planning tool is the use of geographical maps of high resolution [22]. The map is needed to calculate the path loss prediction and subsequently the link loss data. For network planning purposes, a digital map should include at least topographic data (terrain height), morphographic (terrain type, clutter type), and building location in the form of raster maps. Therefore, when planning a 3G system, a resolution of at least 5m is required for dense urban areas, since geographical cell sizes will be rather small.

The next planning step after the site location selection is the optimization of the dominance areas for each cell. In this context, dominance is related only to the propagation conditions. Dominance area optimization is crucial for interference and soft handover area, as well as soft handover probability control. Naturally, since traffic is not distributed uniformly and propagation conditions vary, the cell dominance areas can never be exactly evenly distributed and also vary in size.

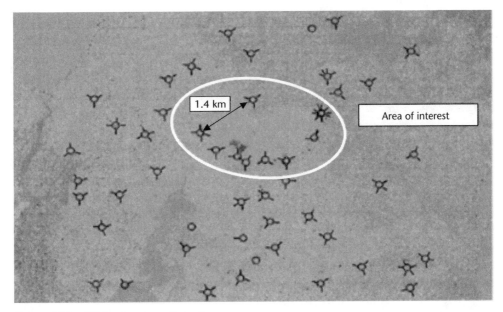

Figure 4.11 GSM network configuration.

4.2.2.1 Traffic Profiles

Traffic modeling and service requirements form a basis for advanced network planning and for evaluating the interaction of coverage and capacity. The more accurate the traffic profile is estimated, the more realistic the results will be. In the traffic modeling phase, traffic forecasts are created in different ways. The busy-hour traffic can be given as input figure, or measured traffic data from a measurement tool. For example, knowledge of hot-spot locations in the current network and traffic measurements from these locations is useful. Therefore, a radio network planning tool has to import traffic information from GSM network measurements, since hotspots are often located in the same area independent from the radio access technology or method.

4.2.2.2 Simulation Planning Scenarios and Details

The planning presented in this section has been based on the 8 p.m. traffic profile, since it is considered the busy hour of the day. The selected UMTS planning configurations are described below as three different scenarios. In the first scenario depicted in Figure 4.12, two three-sectored sites are planned on the existing planned positions, while in the second scenario three single sectors are planned on the initial GSM planning. Finally, three-sectored UMTS sites at optimum positions have been selected in scenario 3.

Several levels of penetration over the GSM profiles have been examined, as well as service distribution alternatives. Alternative mixtures for the indoor/outdoor/in-car users' percentage are displayed in the following simulation analysis.

As far as the simulation characteristics are concerned, these are configured as follows: wideband power amplifiers (WPAs) of 20W/carrier and low noise amplifies (LNAs) to all sectors are assumed. Two-way diversity instead of Tx diversity is

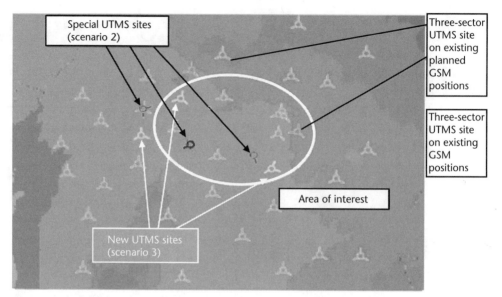

Figure 4.12 UMTS simulation sites.

considered, and soft handover is allowed to all services. The coverage files, traffic profile and GSM cell dominance resolution is 5 × 5m.

To measure the capabilities of the network, the class of algorithms used are based on Monte Carlo methods, which is based on assumptions about performance and/or the link failures. A trade-off between time and accuracy is taking place when using these methods. The more time is used, the better the results of the estimation. In this Monte Carlo simulation, a pixel resolution of 30 × 30m and a statistics generation of 5 occupancies/pixel of PS 64/64 is obtained.

4.2.2.3 Analysis of Different Simulation Scenarios

An analysis of the outcomes of the three different simulation scenarios follows. The initial assumptions taken into consideration are the following: The traffic profile is based in the observations made on the 8 P.M. case. Moreover, UMTS penetration over GSM traffic is assumed to be 10%, while the services are distributed 50% for voice, 35% for PS 64/64, and 15% for PS 64/128. Furthermore, the user's profile is assumed to consist of 80% in-door, 18% outdoor, and 2% in-car and finally, it has been decided that no RRM algorithms are to be used.

4.2.2.3.1 Scenario 1

In the first scenario where two three-sectored sites are planned as it is depicted on Figure 4.13, it is shown that there are some areas of low coverage probability, stretching from top left to bottom right of the planned area. These areas represent a coverage below 50% and will be the area of focus through the planning process. As far as the remaining area is concerned, the coverage probability is certainly above 90% with a few spots of 80% or even 70%, which could be characterized to have acceptable performance as they are.

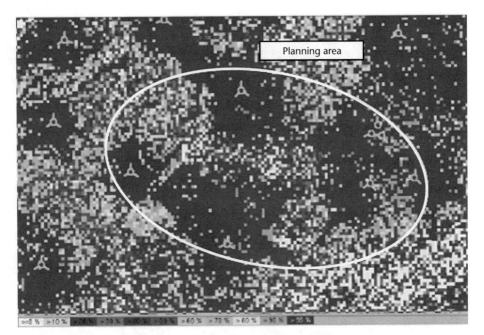

Planning area

Figure 4.13   Coverage probability of selected GSM sites.

4.2.2.3.2 Scenario 2

In the second scenario, where special UMTS sites are introduced (Figure 4.14) a noticeable improvement has been achieved, but it is still well under the desired target regarding the required sufficient coverage probability. It could be assumed that the rise in coverage is about 10% (i.e., 60% on average in the problematic areas) but is below 90%, which is the optimum coverage threshold.

4.2.2.3.3 Scenario 3

In the third scenario, presented in Figure 4.15, where three additional UMTS sites are planned in optimum positions, it is evident that a large improvement has been accomplished. The top left area has reached optimum coverage probabilities, while the bottom right has been reduced significantly. In the first case, an increase of about 30% is achieved, while in the second the coverage probability reaches 100% in some spots and in the remaining areas varies between 50% and 70%.

It is obvious that the third scenario represents the most satisfying results, since the coverage probabilities reach best percentages for the majority of the areas of interest. This will therefore be selected as the more feasible scenario and will be the focus of the planning process.

A further analysis could be performed on the three different scenarios with respect to the outdoor, in-car and in-door probabilities separately. Figure 4.16 depicts the probabilities for the first scenario. The outdoor probability is optimal with very few problematic areas. The in-car probability is clearly lower than the outdoor averaging at about 80%. As far as the indoor probability is concerned, the coverage is drastically reduced and falls below any acceptable levels in the majority of the area of interest.

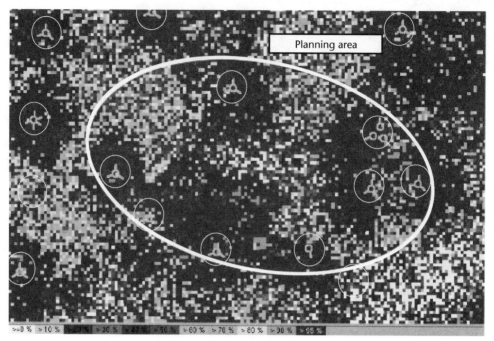

Figure 4.14   Coverage probability of GSM sites with UMTS sectors.

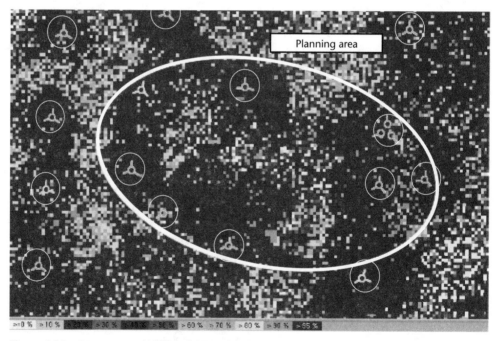

Figure 4.15   Coverage probability of GSM plus three UMTS sites.

For the second scenario, the following observations could be made. The outdoor coverage remains optimum, while for the in-car coverage there is a small improvement

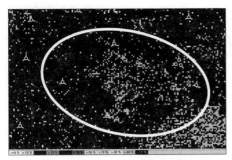

Outdoor coverage probability

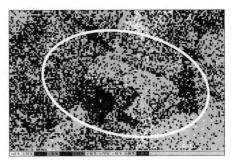

In-car coverageprobability

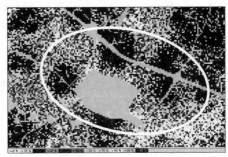

Indoor coverage probability

Figure 4.16 GSM sites coverage probabilities for different areas.

of about 5% on average in the overall probability. A similar improvement is achieved in the case of the indoor coverage but still far from an acceptable probability, as depicted in Figure 4.17.

As it is expected the third scenario presents the most satisfying results. For the outdoor and indoor coverage the percentage remains at an acceptable level but in the case of the indoor the coverage changes drastically. The problematic areas in the circumference of the area of interest reach the optimal coverage percentage, while the remaining problematic area increases in an average of 10% in comparison to scenario 2 reaching a coverage of about 70%, as depicted in Figure 4.18.

4.2.2.4 Performance for Variable Distribution

The performance characteristics for two different distributions of outdoor/indoor/in-car users for scenario 3 are tabulated in Table 4.2. The load factor is 50%, and the penetration of UMTS over GSM traffic is assumed to be 10%. Five performance characteristics have been selected: service UL/DL, UL power outage, load factor outage, average UL throughput, and average DL throughput.

The decisive performance characteristic is the uplink power outage since the other four parameters do not change significantly in the two distribution scenarios. The power outage is at acceptable percentages in the second case, and therefore it can be assumed that by allowing more outdoor and in-car users the UL power outage is improved.

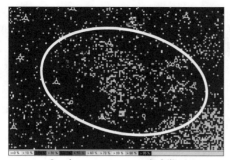
Outdoor coverage probability

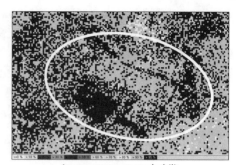
In-car coverage probability

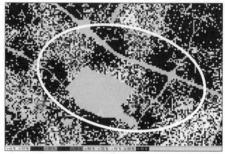
Indoor coverage probability

Figure 4.17   GSM sites with UMTS sectors' coverage probabilities.

4.2.2.5   Hourly and Average Area Performance

Table 4.3 displays the hourly performance for the planning scenarios 2 and 3. The performance characteristics examined are the UL power outage, the average UL throughput and the average DL throughput. The load factor is assumed to be 50% while a penetration of 10% over GSM traffic is selected.

It can be observed that for scenario 2, the only acceptable performance is monitored at 9 A.M., since for the remaining three cases the power outage for PS 64 and PS 128 exceeds 10%, which is not considered a feasible situation. Similarly, for scenario 3 the only problematic performance appears at 8 P.M., while for all other cases it is within the acceptable percentages.

On the second table, the average performance for the simulated hours is displayed for both scenarios. Scenario 3 is clearly the acceptable one, since it achieves percentages over 90% for all three services, while in the case of scenario 2 the two packet services fail to reach the threshold.

4.2.2.6   UMTS Service Distribution Impact

In Table 4.4, the performance for various service distributions has been examined for the worst case of the third scenario (8 P.M.), since it has been considered the most acceptable of all three presented here. The load factor is selected again to be 50%, and the penetration of UMTS is considered to be 10%. Three distribution scenarios have been considered and the performance indicators are examined for each one of them.

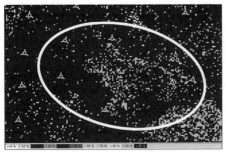

Outdoor coverage probability

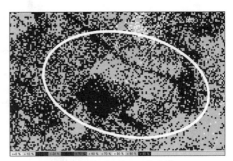

In-car coverage probability

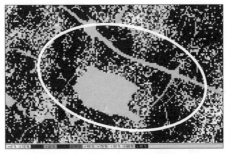

Indoor coverage probability

Figure 4.18   GSM plus three UMTS sites' coverage probabilities.

Table 4.2   Performance for Variable Distribution

Performance for Different "Outdoor/Indoor/In-Car" Distribution
(*Load Factor 50%, 10% over GSM Traffic)
Scenario 3

Outdoor/Indoor/In-Car	Service UL/DL	UL Power Outage (%)	Load Factor Outage (%)	Average UL Throughput (Kbps)	Average DL Throughput (Kbps)
18%/80%/2%	Voice 12.2	4.9	0.5	1,042	1,320
	PS 64/64	10.9	1.5		
	PS 64/128	11.3	1.5		
30%/60%/10%	Voice 12.2	4.0	0.5	1,060	1,346
	PS 64/64	9.0	1.7		
	PS 64/128	9.0	1.7		

In the first case a distribution of 50% voice, 35% PS 64, and 15% PS 128 is considered. In this case the load factor lies below 2%, which is the threshold for the percentage and thus it is an acceptable case. In the next two cases where voice falls below 40% of the overall distribution it is observed that the load factor exceeds 2% significantly, which demonstrates that these cases cannot be accepted.

It is obvious that the dimensioning and planning of a UMTS network is much more complicated compared to the one of a 2G network. On the other hand, the above simulations prove that based on the existing planning, the operator can achieve adequate coverage, by providing only a few additional sites. This of course requires accurate planning tools and experienced network designers.

Table 4.3 Hourly Average Performance

Hourly Performance for the Planning Scenarios (Load Factor 50%, 10% over GSM Traffic)

	Scenario 2				Scenario 3		
Hour	Average UL Throughput (Kbps)	UL power Outage (%)	Average UL Throughput (Kbps)	Average DL Throughput (Kbps)	UL Power Outage (%)	Average UL Throughput (Kbps)	Average DL Throughput (Kbps)
9 A.M.	Voice 12.2	3.1	699	888	3.4	752	951
	PS 64/64	9.2			9.0		
	PS 64/128	9.3			8.4		
1 P.M.	Voice 12.2	3.4	924	1,172	4.0	996	1,268
	PS 64/64	10.6			9.8		
	PS 64/128	10.6			9.7		
4 P.M.	Voice 12.2	3.3	740	941	3.4	794	1,004
	PS 64/64	10.1			8.8		
	PS 64/128	10.4			9.0		
8 P.M.	Voice 12.2	4.3	960	1,281	4.9	1,043	1,320
	PS 64/64	12.5			10.9		
	PS 64/128	12.5			11.3		

Average Performance for the Planning Scenarios

Service	Scenario 2 (%)	Scenario 3 (%)
Voice 12.2	96.4	96.0
PS 64/64	89.3	90.5
PS 64/128	89.2	90.3

Table 4.4 Service Distribution Impact

Performance for Various Services Distribution on GSM Traffic (Load Factor 50%, 10% over GSM Traffic)

Scenario 3

UMTS Services Distribution	Service UL/DL (Kbps)	UL Power Outage (%)	Load Factor Outage (%)	Average UL Throughput (Kbps)	Average DL Throughput (Kbps)
50%	Voice 12.2	4.9	0.5	1,043	1,320
35%	PS 64/64	10.9	1.5		
15%	PS 64/128	11.8	1.5		
35%	Voice 12.2	4.9	1.0	1,263	1,625
45%	PS 64/64	10.8	3.0		
20%	PS 64/128	11.1	2.9		
20%	Voice 12.2	4.9	1.8	1,459	1,896
55%	PS 64/64	11.2	4.5		
25%	PS 64/128	11.2	4.5		

4.3   Resource Management Techniques and Guidelines for Implementation

Various radio resource management techniques for the 3G UMTS network are presented in this section. The support for RRM procedures from 3GPP specifications is presented, and the constraints implied by them are given. In most cases, references are made to papers explicitly addressing WCDMA or UMTS, but also CDMA papers are sometimes taken into account if the information is almost directly applicable to UMTS. Attention is mainly focused on FDD mode of UMTS because it will most likely be deployed first. However, TDD mode resource management techniques are also briefly addressed at the end of this chapter.

   In the first part, issues affecting resource management in UMTS are described. The impact of different service types, channels, network structure and user behavior on RRM is explained. After that, the main categories of RMTs, power control, handover, admission control, load control, and scheduling, are described from the point of view of 3GPP specifications and recent research papers. Finally, miscellaneous techniques, TDD mode RRM and inter-RAT (UMTS/GSM) RRM is investigated.

4.3.1   Issues Affecting Resource Management in UMTS

In the following sections, the various issues that affect resource management in a UMTS network are presented. This is quite important for the better understanding of each resource management technique that is presented afterward.

4.3.1.1   Different UMTS Releases and Modes

As noted in the beginning of this chapter, there are different releases of UMTS specification available, and the structure of the UMTS network differs quite significantly between releases. Release 99 uses the GSM/GPRS core network structure but incorporates the new radio access technology part (UTRAN). Also, new features such as high-speed downlink packet access (HSDPA) are being introduced in the more recent releases (Release 4, Release 5). All of these things will unavoidably have some effect on the management of UMTS network resources. For example, HSDPA brings with it the possibility for carrying out fast scheduling in node B compared to scheduling in RNC.

   From the two modes of operation (FDD and TDD) specified for UMTS, TDD mode offers more flexibility in resource allocation for asymmetrical service types. However, the FDD mode is expected to be deployed first and to provide most of the capacity, and thus this analysis concerning resource management techniques is mainly focused on FDD mode techniques. A short presentation of TDD mode techniques can also be found at the end of this chapter.

4.3.1.2   Limitations of Network Performance

We know that the air interface is the main bottleneck of cellular networks. The available frequency band is a limited resource, and thus its utilization should be optimized. The capacity of CDMA-based networks, such as UMTS, is limited by the interference that the users can tolerate. In principle, all node Bs operate on the same frequency band and thus transmissions cause interference to each other. Of course, if additional capacity is required, it is possible to resort to multiple carrier frequencies as well.

   In the initial network planning phase, locations for node Bs are selected, and suitable values for network parameters (e.g., maximum transmission powers) are determined. The network plan is based on estimated offered traffic and its spatial distribution. However, the offered traffic does not remain constant in time but exhibits changes. Using suitable resource management techniques, it is possible to take into account changing load situations after the deployment of the network and balance the traffic to some extent.

4.3.1.3   CDMA-Related Features

The radio access technology used in UMTS networks, namely WCDMA, has some features that are relevant from the point of view of resource management. One such feature is the cell breathing phenomenon. This means that the cell coverage area depends on current network load situation. When the offered traffic is increased, the interference increases and UEs are required to use higher transmission powers to attain sufficient call quality. However, due to the limitations on transmission power, remote UEs cannot attain the required SIR even using maximum available transmission power. Thus, the cell coverage area decreases.

Another issue is the soft capacity: adjacent cells cause interference to the central cell and thus the capacity of the central cell is dependent on the load situation of the adjacent cells as well. If adjacent cells have low load, more users can be served in the central cell than when the adjacent cells have high load.

Cell breathing affects resource management in the way that it causes unfair conditions between users depending on their distance from the node B. Soft capacity indicates that it is essential to manage not only the resources of the target cell but also of the adjacent cells.

4.3.1.4   Service Types and QoS Levels

In addition to normal speech, other service types such as video and WWW browsing are expected to be used in UMTS networks. They have different QoS requirements (e.g., BER, max delay, and delay jitter) and different resource management techniques can be applied. Nonreal-time services tolerate some delay, and they can be more freely controlled by scheduling than real-time services. Furthermore, service types can have multiple QoS levels associated to them enabling QoS negotiation and QoS renegotiation procedures. In that way, it is possible to assign a lower QoS level than the user has requested in call admission or downgrade the QoS levels of ongoing calls if the network is congested. Furthermore, it is possible to have multiple services at the same time active in one terminal, and the data streams can be controlled and prioritized in MAC layer using scheduling procedures.

4.3.1.5   Different Channels

A diverse selection of available service types requires different channels. As we have seen in Section 4.3.3. the transport channels available for data transmission are the dedicated channel (DCH, UL, and DL), random access channel (RACH and UL), common packet channel (CPCH and UL), forward access channel (FACH and DL), and downlink shared channel (DSCH and DL).

Dedicated channels are most suitable for transmission of speech. From the point of view of resource management, they enable the usage of inner-loop power control and soft handover. The downlink shared channel is suitable for transmission of bursty nonreal-time data and it avoids wasting orthogonal variable spreading factor (OVSF) code resources. Efficient methods for scheduling the transmissions of multiple users are required.

4.3.1.6   Network Structure

In the network deployment, decisions about the structure of the network have to be made. Multiple cell layers (micro/macro) could be used to avoid frequent handovers between microcells, and multiple carriers could be used in one node B to increase the capacity. Both of these affect resource management in relation to the handover algorithm: There has to be some way of deciding which users are assigned to which cell layer (e.g., based on estimated user speed). Furthermore, if multiple scrambling codes are used in one cell, an efficient way of assigning users to either scrambling code has to be devised. For each cell, the neighboring cell list has to be defined appropriately because handovers can occur only to those cells.

4.3.1.7   User Behavior

User distribution in the area in which the network is deployed affects the network load per cell. In exceptional situations, the spatial distribution of offered traffic might differ significantly from the initially estimated distribution, and this could cause overload to some cells of the network. Resource management techniques could be used to attempt to balance and redistribute the traffic between cells in such traffic load situations.

User mobility causes constant changes in channel conditions and efficient power control algorithms are required to maintain transmission powers at a level that ensures requested QoS for the call. Handovers are affected by user movement as well and fast moving users undergo more handover operations than slowly moving users. In the handover algorithm, too frequent handover operations should be avoided (ping-pong effect) in order to prevent excessive signaling load.

An issue affecting the design of resource management techniques that block or drop calls in a controlled manner is the propensity of users to retry call establishment if they are not given access to network or their calls are terminated prematurely. This causes additional load to the network.

4.3.1.8   Transmission Directions: Uplink and Downlink

The traffic load in the downlink direction is expected to be higher than in the uplink direction in UMTS networks due to the usage of asymmetric service types (e.g., WWW browsing). Thus, in the resource management, this fact should be taken into account and efficient management of resources in the downlink direction should exploit for example the characteristics of nonreal-time services by scheduling them.

In general, node B can more efficiently control resources in the downlink direction because it has knowledge of all the users. In the uplink direction, measurements made by UE first have then to be reported to the network. Also, the HSDPA feature includes fast scheduling in the downlink direction to be carried out on the high-speed DSCH (HS-DSCH) which is a new shared downlink transport channel terminated at node B.

Some resource management techniques have different impacts in the downlink and uplink directions. In soft handover, the combining schemes used can be different (maximum ratio/selection combining) and differences in gain exist between the uplink and downlink. Also, site selection diversity transmission (SSDT) mode in soft handover is targeted to have an impact on the downlink direction. The capabilities of node B and UE in terms of maximum transmission power often differ: More total power is available in node B but, on the other hand, it has to be shared among multiple users. Thus, this affects the management of the power resource.

4.3.2 Power Control

The key feature of any CDMA system is the fact that all users of the system share common power/interference resource. Spreading of signal makes it possible to decrease it below noise level (the difference between the required narrowband signal level and its level while still spread is called processing gain and depends on spreading factor). In documents about UMTS the relative level of spread signal compared to the interference is called *SIR*; it is usually expressed in decibels. Since processing gain is fixed for a certain channel and the signal after despreading must have a certain quality (expressed often as E_b/N_0, bit energy per noise), the minimum *SIR* level is known at the receiver and is expressed as SIR_{target}. The actual *SIR* must not be lower than SIR_{target}, but on the other hand a too high *SIR* would waste resources (available interference level). Furthermore, the overly high *SIR* of a user located nearer suppresses completely the signal from other users, located farther (near-far effect). Thus *SIR* depends on the power of received signal and momentary interference at the receiver. To enable a transmitter to provide a signal received at a proper level in an environment that can vary a lot it is necessary to allow it to change the transmission power.

The mechanism controlling those changes must be able to derive proper transmission power, either by means of measurements and internal calculations, or by a feedback loop, and react quickly enough to changes in the environment. The dynamism of the mechanism depends on tackled problems: noise level, fading, and other characteristics of the signal and the environment. The mechanism is called power control (PC). Figure 4.19 shows how varying TX power of the transmitter of a moving UE enables maintaining stable *SIR* (except short *SIR* drop due to sudden interference change; the plot does not include fast fading).

In UMTS (FDD) there are four types of PCs [23]:

1. *Inner loop:* Fast power control performed on all data channels (except PRACH and CCPCH). This requires a feedback information in a reverse channel.
2. *Outer loop:* Slow power control, enabling changing of SIR_{target} to reflect changes in the channel characteristic. This is performed internally in the receiver based on measurements and calculations.
3. *Closed loop:* Fast power balance between transmitters when transmission diversity is used (e.g., in soft handover). It requires feedback information from the receiver.
4. *Open loop:* Used to calculate initial power of a transmitter when a new radio link is to be established. This is based on internal measurements and calculations.

The idea of PC is presented in Figure 4.20.

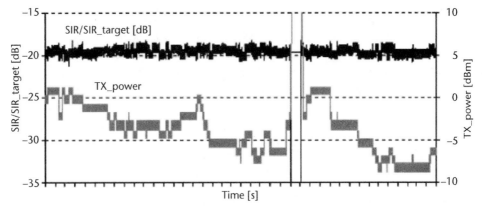

Figure 4.19   Stable *SIR* versus varying TX power.

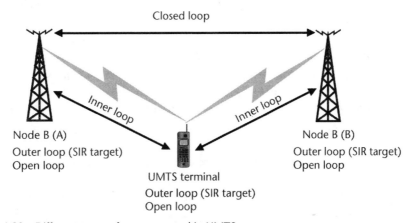

Figure 4.20   Different types of power control in UMTS.

4.3.2.1   3GPP Constraints/Support

The power control mechanism is well specified in the 3GPP UMTS standards. This is necessary because of the loop-based algorithms, which must work in the same way with the equipment coming from all vendors.

The standardization starts by defining requirements for radio transmission and reception [24]. All mobile terminals must be able to lower their power to less than −50 dBm, and increase it to the upper limit precisely defined for each UE class (four classes: 21, 24, 27, and 33 dBm). Similarly, other characteristics of transceivers and receivers are defined as minimum requirements. Due to fast fading it was decided that fast power control should be performed 1,500 times per second. Additionally the standards present test procedures and parameters.

The actual power control mechanism is presented in [23]. Inner and outer loops for power control are used in dedicated and shared channels (those that enable transmission longer than 1–2 packets). For dedicated channels, in case of soft handover closed loop is necessary for downlink (power balance between node Bs). Open loop is required for all channels to initialize transmission.

The uplink is more important because of noncoordinated transmission from mobile terminals. A node B compares momentary SIR to SIR_{target} and sends to a UE a transmit power control (TPC) command ("up" or "down"). UE must unconditionally follow the received commands as long as its transmitter operates within the available power range. There are two algorithms defining UE's behavior upon reception of a TPC command: The first one just defines immediate action, whereas the other takes into account the last five TPC commands. As the result UE's power increases, decreases (power step is either 1 or 2 dB), or remains at the same level (this is possible only with the second algorithm). When a UE is in soft handover state it receives several commands, which must be combined. The standard defines requirements for the combining algorithm, but not the actual procedure. If the side selection diversity technique (SSDT) is used, UE follows TPC commands from the primary link. The algorithm that is to be used and the power step are decided in the network (the procedure is not specified).

In the downlink the resources are controlled centrally, from a node B. That enables implementation of interference control techniques, like OVSF codes, and a more liberal specification. A node B receives TPC commands from UE, but it is not defined how a UE derives a TPC command (a similar algorithm to the one defined for the downlink is proposed). The specified algorithm for a node B enables the C-RNC to influence the final output power change and therefore leaves a space for special RRM techniques. In case of soft handover, when SSDT is not used, UE combines signals received from all links, which enables the nodes B involved in the transmission to use lower powers. They all receive the same TPC commands, but UE informs them also about the quality of each link, which is necessary for closed loop power control. To enable constant power control on shared channels there must always be allocated a dedicated channel (only control part) in the reverse direction.

The algorithm to derive SIR_{target} (outer loop power control, which is the RRC function [25]) is not specified, but it is in both directions controlled from the network SIR_{target} and depends on measurements of BER and BLER to maintain required connection quality.

The initial power in uplink transmission (all channels) is clearly defined (may be partially based on measurements), but always controlled from the network [25]. In case of DCH the power is set according to UE measurements and a parameter sent from the network. In case of RACH and CPCH a special power-ramping mechanism is defined: UE sends consequent access requests, each with higher power. The final transmission power is set according to the last (acknowledged) request. The power of the first request is calculated according to UE measurements and information received from the network (it contains both network measurements and an additional parameter enabling usage of special RRM techniques). In the downlink the algorithms for open loop are not defined.

4.3.2.2 Evaluation Criteria

Since the objective of power control is very clear—to provide required reception quality with optimal resource usage—the criterion to evaluate the PC algorithms is to try to assess their efficiency. They vary, however, with different indicators used as the efficiency indicators. Most authors propose mean transmission power [26, 27].

In this case it is assumed that in the same simulation conditions the better algorithm will result in lower transmission power in the analyzed direction. This may be amended, and the mean power can be calculated per user (in downlink) [27]. Another, more general approach is to use a popular indicator: outage probability [27]. Finally in some documents capacity is used: either as a number of users, which is similar to the outage indicator, or as "data capacity" (Kbps/MHz/cell).

4.3.2.3   Algorithms

The basic UMTS power control algorithm is quite straightforward and simple. Currently research activities focus mostly on soft handover and SSDT cases. It is predicted [28] that in the soft handover state the downlink would be higher loaded as compared to hard handover. One possibility for enhancing the proposed power control, but not violating existing standardization, consists of dynamic switching between normal power control with power balancing (closed power control loop) and SSDT in the soft handover state [27]. As long as there is enough power available at a node B a classic power control is used. However, when the upper limit of the transceiver is getting near all connections are switched to SSDT mode. According to presented results, this enables more economic power utilization at the cost of lower quality of affected calls.

Signal attenuation could be used to balance power between node Bs involved in soft handover [26]. The transmission power of a node B assigned for the connection to the UE is either calculated directly from the relative attenuation measured at the UE (the bigger loss the higher TX power), or the measurement is used to adjust SIR_{target} in UE and then the power of a node B is adjusted in the classic way according to the modified SIR_{target}. This approach enables decreasing the mean TX power at a node B still maintaining the gain of soft handover. However, the impact of constant measurements that must be performed at a UE and then transmitted to the network is unknown. The latter solution requires also sending separate TPC commands to each node B, which is not supported in 3GPP standards. Another method of closed loop requires coordination from C-RNC [29] It sends to all nodes-B involved in soft handover a reference power and a coefficient-defining importance of the reference power. This is taken into account when new TX power is calculated at a node B. The balancing helps control power drifting and therefore decreases the average TX power of a node B. The technique is completely compliant with the standards.

The actual power control algorithm can be amended, too. Instead of sending only binary TPC command, the actual difference between SIR and SIR_{target} (or an approximation of the difference) could be transmitted [30]. In this case the power control step would be calculated according to the difference. Such a dynamic power control step enables much quicker convergence when radio condition changes rapidly (e.g., the interference increases). Another algorithm, also noncompliant with the standards concerns SSDT. It proposes enabling usage of multiple primary links, instead of just one [31]. A UE in soft handover state could select out of the whole active set the best links, if the difference between them is small. This would increase connection reliability without increasing downlink interference caused by significant differences in signal strength among links in the active set.

4.3.2.4   Conclusions

The research concerning the power control in UMTS allows us to point to some possible sources of problems: first is the reliability of a received TPC command—commands interpreted incorrectly due to bad signal quality may lead to significant power drift and hence a need to transmit the TPC part of a message with a higher power. In the case of soft handover the drift affects not only one cell, but all involved in the procedure. Here, an additional possibility is a good closed loop power control mechanism [29]. Finally SSDT and normal power control with balancing in soft handover were compared [27]. Results show that SSDT gives better results, when, due to network configuration, relatively many users are in soft handover state (e.g., in macrocells), whereas when soft handover occurs seldom (e.g., in microcells) normal power control offers better performance.

4.3.3   Handover

Handover is a procedure that has been known since the fist cellular networks were designed. It enables a user who has an ongoing call and is moving to be handed over from his or her current cell to one of the adjacent cells. Therefore handover enables users' mobility. In CDMA-based networks, there are two general types of handover:

1. *Hard handover (HHO):* Performed when a UE cannot communicate with more than one base station at a time because of technical conditions. In UMTS this is the case when the cells use different frequency band. During HHO a user's terminal must first break the connection to the current cell, reconfigure its transmitter, and try to establish a connection to the new cell.
2. *Soft handover (SHO):* Performed when a mobile terminal can communicate with several base stations at the same time. In that case a terminal first establishes the new link and then drops the old one. In some cases a terminal may keep several links active. The set of active cells at a certain time is called an active set (AS).

In UMTS a UE can operate only at one frequency at a time, so the hard handover occurs always when adjacent cells use different frequencies. This may happen within the UMTS network, but it always happens when the cells represent different access technologies (e.g., FDD/TDD, or UMTS/GSM) as it is intersystem handover. When the cells belong to the same access technology and operate on the same frequency a soft handover can occur. It not only improves reliability of a connection ("first establish, then drop"), but also allows for control of uplink interference when the terminal is near the cells' boundary (the terminal can rely on the better link, as described in the power control section and therefore use less power). A special case is when a UE has in its active set two sectors of the same site. This is called softer handover and enables signal combining at the site. A drawback of a soft handover is higher resource allocation in the downlink. In some special cases soft handover is not allowed, though it is technically possible—e.g., when a user has allocated a downlink DCH as a feedback channel for CPCH.

A handover, either hard or soft, is always controlled from the network—the involved terminal provides the network with measurements, but the decision is made in the RNC. A handover may occur only when a terminal has a dedicated resources assigned (in CELL_DCH RRC state); otherwise cell reselection is performed (decision of cell reselection is made in the terminal).

4.3.3.1   3GPP Constraints/Support

Since handover is controlled from the network the actual algorithm to make the decision on the link addition or drop is not standardized. This is left as an area for research and a field for equipment vendors to compete. The handover procedures, however—link addition, drop and replacement—as well as measurements and performance requirements are specified in the 3GPP documents very precisely.

The measurements performed by a UE, more important for a handover, can be reported in three ways: periodically all the time, periodically only after a certain condition is fulfilled, or only at the moment the condition is fulfilled [32]. The method depends on the network (a UE must be able to operate in all ways) and is sent to a UE with the `MEASUREMENT_CONTROL RRC` message [25]. Continuous periodic reporting is not required in most situations and consumes resources (causes interference). Usually it is enough, that RNC establishes certain conditions and a UE reports only at the moment the condition is fulfilled. Later the periodical reporting is possible as long as the condition is valid, or other condition is fulfilled. A network can very precisely define what is to be reported, how often, and what should be measured (e.g., only active set cells, neighboring cells, or all known cells). Events of three groups can be utilized for handover:

Intrafrequency reporting events for FDD are used to monitor cells belonging to the same frequency (including active set cells). They are listed as follows.

- *Reporting event 1A:* A primary (common pilot channel) CPICH enters the reporting range;
- *Reporting event 1B:* A primary CPICH leaves the reporting range;
- *Reporting event 1C:* A nonactive primary CPICH becomes better than an active primary CPICH;
- *Reporting event 1D:* Change of best cell;
- *Reporting event 1E:* A primary CPICH becomes better than an absolute threshold;
- *Reporting event 1F:* A primary CPICH becomes worse than an absolute threshold.

Interfrequency reporting events, listed in the following, are used to monitor other frequency bands of UTRAN.

- *Reporting event 2A:* Change of best frequency;
- *Reporting event 2B:* The estimated quality of the currently used frequency is below a certain threshold and the estimated quality of a nonused frequency is above a certain threshold;
- *Reporting event 2C:* The estimated quality of a nonused frequency is above a certain threshold;

- *Reporting event 2D:* The estimated quality of the currently used frequency is below a certain threshold;
- *Reporting event 2E:* The estimated quality of a nonused frequency is below a certain threshold;
- *Reporting event 2F:* The estimated quality of the currently used frequency is above a certain threshold.

Inter-RAT reporting events, listed as follows, are used to monitor cells belonging to other systems or access technologies and therefore enables intersystem and interRAT handovers.

- *Reporting event 3A:* The estimated quality of the currently used UTRAN frequency is below a certain threshold, and the estimated quality of the other system is above a certain threshold;
- *Reporting event 3B:* The estimated quality of other system is below a certain threshold;
- *Reporting event 3C:* The estimated quality of other system is above a certain threshold;
- *Reporting event 3D:* Change of best cell in other system.

Requirements for a UE for the measurements (delay, precision, and selectivity of the receiver) and a terminal's capabilities related to handover are standardized [e.g., interfrequency measurement must not exceed the gap of the compressed mode (i.e., a transmission mode when a terminal changes its operation frequency for a short period of time to perform measurements of another frequency)]; a new cell must be detected within 800 ms; a UMTS terminal must be able to bear at least 6 links during the soft handover state [25, 33].

Each type of a handover is initiated with a RRC command sent from the network [25]. SHO within UTRAN is initiated with `ACTIVE_SET_UPDATE` (the message can be used for link addition, removal, or combine several additions and removals), HHO uses one of the reconfiguration messages:
(`PHYSICAL_CHANNEL_RECONFIGURATION`, `TRANSPORT_CHANNEL_RECONFIGURATION`, `RADIO_BEARER_RECONFIGURATION` Alternatively, `RADIO_BEARER_ESTABLISHMENT`), inter-RAT handovers use `HANDOVER_TO_UTRAN_COMMAND` (if the handover is from UMTS to other system: GSM or cdma2000) or `HANDOVER_FROM_UTRAN_COMMAND` (as a handover to UMTS). Examples of different handovers (message flow) can be found in 3GPP documents.

Usage of hierarchical cell structure (HCS) for handover purposes enables an operator to divide its radio network into layers and to provide different types of coverage (indoor, microcells, macro/umbrella cells). It lies beyond the scope of the 3GPP standards (again, because the handover algorithm is not standardized). UE must be, however, aware of HCS's existence to correctly perform cell selection and then reselections. Therefore the HCS information is present in system information blocks (SIBs) 3/4 and 11/12 broadcast in each UMTS cell [25]. It enables usage up to 8 HCS levels. In case of handover, HCS will rather affect HHO, since the layers are usually implemented at different frequency bands.

4.3.3.2   Evaluation Criteria

To assess handover performance several indicators have been proposed. The first group aims directly at the procedure: A number of HO operations is used (calculated as a mean per call, total during a simulation or some period of time, a probability of HO or finally HO rate) [34, 35]. This approach takes into account the fact that each handover operation causes additional interference due to an exchange of the signaling messages. Furthermore, a handover interrupts data transmission, which either degrades call quality or requires additional mechanisms to tackle the problem. The same issues are taken into account when "ping-pong" handovers (i.e., unnecessary handover operations) are taking place [34]. Link holding time (the mean time a link is kept in the active set) is used: The longer the time is the less often a handover is performed.

Another, more general indicator may be system capacity (total, or only in one direction). This is quite straightforward—the higher the capacity the better the algorithm. Beside capacity is calculated as number of users, in a multiservice environment data capacity can be calculated (Kbps/MHz/cell) [36]. Capacity is related to call blocking (regarded as unsuccessful handover) and outage probabilities and affected by signal quality (bad quality may lead to premature call termination) [34]. Sometimes a simple check, to verify only whether an algorithm can provide good enough signal quality is enough [37].

Finally there are some specialized indicators: It may be HO delay (it should not be too long, for then the interference increases significantly, but on the other hand an overly quick decision could lead to ping-pong effect), or a number of candidate cells available when the HO decision is to be made (if there are more candidates, the probability of HO block is lower) [34].

In case of soft handover, the gain may be calculated as compared to the HHO case. Most of the presented criteria may be used for this purpose. A combination of blocking probability and downlink transmission power can be used; soft handover requires more resources in downlink, but on the other hand it offers better quality in the uplink [38]. Therefore there is a trade-off between the two directions that can be optimized. Another indicator, also usable only to compare SHO algorithm is SHO surplus—soft handover always requires more resources to be involved in a connection, so there is a surplus as compared to the HHO case. The surplus is the cost of better quality, but different SHO algorithms offering similar quality can be compared for the lowest surplus [39].

4.3.3.3   Algorithms

The simplest possible algorithm, used rather in simulators, is based on users' location. Cell ranges are decided at the network planning stage, and the handover occurs when a user passes the boundary between cells. This can be utilized for both hard and soft handover and equipped with decision delay to avoid the ping-pong effect [36]. In the simple case only distance thresholds are used instead of precise location information. An algorithm like that, though it is simple, is difficult to implement in a real network, for location information is in most cases unavailable and in the rest imprecise.

A practical solution for the lack of location information is substituting the distance with a value representing radio propagation. In CDMA networks based on IS-95 standards there is an algorithm used, where a links is added to the active set, when pilot channel strength (measured at a mobile station) is high enough. Similarly, when the power decreases below another threshold the link is dropped. The difference between thresholds and time delay in the decision prevents the algorithm from instability (Figure 4.21). This algorithm is well-known and has been analyzed in many documents [34, 39]. This may be further enhanced—together with the power level criterion based on a comparison of the received pilot channel quality with a combined quality of links already present in the active set [40]. This algorithm cannot be used for hard handover, because absolute threshold would not be stable.

In 3GPP UMTS documents there is no algorithm standardized, but there is one proposed [41]. It is similar to the IS-95 algorithm, but the decision is based on relative thresholds: the best link in the active set is the reference for other links. There are three thresholds: addition, removal, and replacement (used when the active set is full). A new link is added to the active set, when the cell is not worse than the addition thresholds allow as compared to the current best cell. When an active cell becomes worse from the best cell related to the removal threshold, it is removed from the active set. If the active set is full, or for some other reason it should not be enlarged, but there is a cell whose pilot channel is received with a better quality than the worst active cell plus the replacement threshold, the worst cell is replaced with the new one. All the thresholds are supported with a time delay protecting them from decisions based on temporary signal fluctuations. According to the proposal, cell quality is estimated with the received power of its pilot channel. This, however, can be substituted with the SIR of the pilot channel. A simple example of the algorithm is presented in Figure 4.22. This is currently the most popular algorithm, and it has been analyzed in many works [28, 34]. It can be, however, still modified. One possibility, already mentioned, is usage of pilot channel SIR instead of received

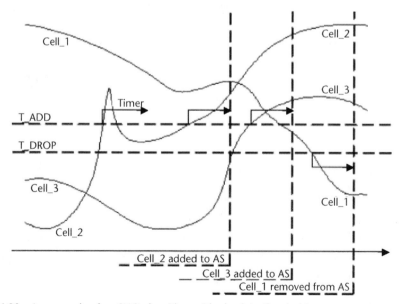

Figure 4.21   An example of an SHO algorithm with absolute thresholds.

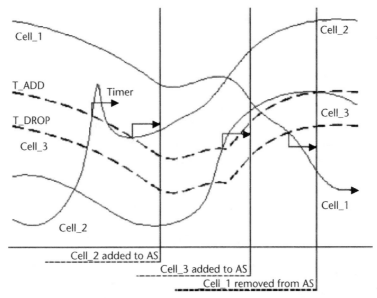

Figure 4.22   An example of an SHO algorithm with relative thresholds (the best cell in the active set is the reference).

power. Furthermore, the algorithm can monitor available downlink resources and limit accordingly the active sets of all, or selected users. This can be achieved either with a modification of threshold values of addition and removal [39], or avoiding adding more links than it is assumed. Variable thresholds can be used also to distribute traffic among cells surrounding a congested cell [42]. Finally, the thresholds can be different depending on a cell, or user's class (this can be a function of the service). The algorithm based on relative thresholds can be used for hard handover, too. In this case there is only one threshold that defines how much a new cell must be better than the current one to have the user handed over.

The pilot channel plays an important role in the handover; therefore it can be used to alter handover operations: A cell transmitting its pilot channel with higher power than its neighbors will attract more users. This phenomenon can be used to balance unevenly distributed traffic. A practical solution could be based on constant load monitoring in all cells and adequate modifying transmission power of their pilot channels to keep the load in each cell at the optimal level [43]. However, the cell breathing effect (related to cell load) must be considered.

The HCS can be also used for handover purposes. A simple algorithm assigning users to either of two HCS layers can be based on users' speed [35]. The algorithm must address both idle users (cell reselection procedure) and active ones (handover). The speed may not be always available, but it can be estimated based on location update or handover rate.

4.3.3.4   Conclusions

Soft handover enables more efficient usage of the uplink, because at every moment a UE can rely on the best link [28]. This enables lower power of the transmitter and therefore decreases uplink interference [37]. Lower interference is immediately reflected in either capacity, or general call quality gain. The cost is higher usage of resources in the downlink (transmission power). Furthermore, performed simulations show that in case of nonreal time (NRT) services (packet transmission) SHO enables more efficient and stable downlink power allocation. Additionally, if the SHO thresholds are chosen carefully, it can help decrease transmission delays. Generally, if the computation surplus and additional resources required by SHO are neglected, this type of handover performs better (lower outage). Further improvement in the system capacity is achieved with softer handover. The gain depends also on selected parameters and obtaining expected results may require careful adjustments [28].

One of the most challenging problems for handover in UMTS is unevenly distributed load. On the other hand, it is a quite common situation that occurs always when people gather at some spot. If there are not enough downlink resources to handle the hotspot changing the thresholds, or the power of pilot channels of surrounding cells can help distribute the traffic among them. This solution, however, extends distance and leads to another dangerous situation, when some users using a more remote base station cause severe interference in the base station that was to be unloaded [42].

Handover is one of the most interesting processes in UMTS. Although it seems quite simple, there are many factors involved in it, and therefore its analysis is multidimensional. That is why some issues, like interference caused by HO signaling, impact of cell size and shape (also, user's speed and the direction of movement) on the handover frequency, intermode handover (FDD/TDD) and intersystem handover (e.g., UMTS/GSM) performance are very seldom addressed in research.

4.3.4 Admission Control

Admission control is used in UMTS to decide whether a requested radio access bearer can be established. The objective is to avoid blocking incoming calls in vain but, at the same time to try to maintain the QoS of already admitted ongoing calls. Admission control is especially important in CDMA-based systems, such as UMTS, since the capacity of a cell is not so clearly determined by, for example, the number of available channels as it is in GSM.

Admission control can be triggered, for example, by connection set up, soft/softer/intersystem handover, channel type switching, and code reallocation [44]. The actual admission algorithm can depend on the triggering reason but in all cases, it utilizes measurements of the current network load situation and estimates of the resource consumption of a call to be admitted. As a result of the algorithm, either rejection or acceptance of the request is obtained.

Other radio resource management procedures are closely connected to the admission control. Handover is related to admission control in the way that special consideration is required for handover calls compared to new calls. Load control can be used even after admission control since overload situations can occur due to user mobility and changes in user activity. Furthermore, less stringent admission control can be used for nonreal-time services since packet scheduling can be used to control them after call admission.

4.3.4.1   3GPP Constraints/Support

The admission process resulting in the decision of whether to admit a new user or reject his or her request is not standardized. However the process and tools that may be useful for the purpose are defined at a very detailed level.

A connection is established always from the network side: It is either the MSC server or SGSN (depends on the service to be used) that makes the decision about establishment of radio access bearer (RAB). RAB is the highest bearer in the access stratum and links a terminal directly to the CN. The establishment command (`RAB_ASSIGNMENT_REQUEST` message of RANAP protocol [45]) contains all necessary data (i.e., call type, required resources, and QoS). Since CN is not aware of available radio resources this command is set to a CRNC that controls the area where the terminal is located. The RNC must either set up the connection between the UE and the CN, or report to the network encountered problems (`RAB_ASSIGNMENT_RESPONSE` message of RANAP protocol). The network can then take a proper action (e.g., decrease quality requirements and request the establishment again). If there are available required resources in the radio interface, the CRNC allocates the resources at node B and sends to the UE a command to configure its radio equipment for the connection (`RADIO_BEARER_SETUP` message of RRC protocol [25]; the RRC signaling connection must be established first). When this is done the connection is established and the RNC becomes SRNC.

Analysis of the admission procedure shows that the admission control algorithms must be located in the core network. There, however, the radio resources are not known, so it may be necessary to implement some elements of the algorithm in the UTRAN (RNC).

It is important to mention how a UE selects a cell that it uses to access the network. The procedures of the initial cell selection and later reselections are well standardized. They are made autonomously in a UE when it is switched on and later, when it moves. The network can influence the process with parameters broadcast in each cell, but cannot address a single terminal. When a user decides to initiate a call (or to answer paging) the terminal requests the connection in the cell on which it is camping. The cell reselection and later access procedure may be amended with the following tools:

- *Cell baring:* A cell can be marked as barred (the information is broadcast in the network). In this case no UE [except for operator's terminals and support of localized service area (SoLSA) users—this depends on the type of barring] can select or reselect the cell. The only exception for cell barring is emergency call, if no other cell is available.

- *Access classes:* In the standards there are 16 classes defined: 10 for ordinary users, five for the operator's and public purposes, and one for emergency calls. Each SIM card is randomly assigned to one of the ordinary classes (it may be also assigned to special classes; the emergency assignment is done dynamically when an emergency call is initiated). In a cell it is possible to bar certain classes (one or more). A UE may freely select or reselect such cell, but when a connection is requested it must compare its access classes with the barred list broadcast in the current cell. If all its classes are listed there it must reselect another cell. Since all normal users have exactly one access class assigned and there are

10 such classes, it is possible to decrease requests in a congested cell by even tens of percents. In the random access procedure the access classes are mapped on access priorities to facilitate access for special users (emergency, operator, and public services) [25].

4.3.4.2   Evaluation Criteria

Admission control is roughly about making trade-offs between ongoing calls and new calls. The algorithm should maintain the QoS of ongoing calls and avoid dropping them but at the same time minimize the new call blocking rate. Thus, there are two relevant quantities for assessing admission control performance are dropping probability [46] and blocking probability [47] (or equivalently, accepted traffic [48]). In same studies, they have been combined into one metric in order to facilitate the evaluation of the admission algorithm. This metric is a commonly designated GOS, and defined as a weighted linear combination of dropping and blocking probabilities. Usually, dropping is considered more annoying than blocking and a weight of 10 is given to dropping probability compared to blocking probability [49]. Other weights are also possible and a value of 4 has been used in [50]. The proportion of satisfied users is another way of expressing the efficiency of admission algorithm [51].

Even if the admission of new calls does not lead to the complete dropping of ongoing calls, it might drive them to outage. Thus, outage probability measures the impact caused by the admission of a new call on the QoS of existing calls [47]. Whether the calls residing in outage are dropped or not depends on the call dropping policy used in the network. A metric similar in role to dropping and outage probabilities has been used in [45]. The percentage of admission decisions leading to overload has been taken to indicate the impact that erroneously admitting a new call has on the network load situation.

Blocking of new calls and dropping of existing calls reduces the mean number of users in the system. Thus, this can also be used in evaluating the admission algorithm [47]. In a similar way, the utilization of a cell indicates how well the resources of the cell are reserved for ongoing calls. Finally, to capture the effect of different service types and make them comparable, base station transmission power might be one suitable metric [51].

Various measures for the assessment of admission control algorithms have been used. Most studies indicate the blocking and dropping probabilities for the algorithms or even some combination of them. However, it should be kept in mind that even simple metrics such as dropping probability depend on other factors such as the actual call dropping policy, which could be independent of the admission control algorithm.

4.3.4.3   Algorithms

The admission control should be carried out separately in both directions: uplink and downlink. This is necessary because there might be available resources in the uplink but at the same time the downlink direction might be overloaded by heavily asymmetric WWW service-type users. Furthermore, before the actual admission control algorithm, the availability of physical resources such as downlink OVSF codes should be checked. In most cases, the hard limit of OVSF codes is not expected to be attained, but it has to be verified.

The starting point of admission control algorithms is the usage of some measurements to indicate the momentary network load. Common quantities used are transmitted power in the base station [52] for the downlink direction and received power in base station [46] for the uplink direction. Other measures can also be used (e.g., to attempt to describe the load situation of the nonreal-time users by delays [52] or queue sizes. The simplest conceivable metric is the number of users [48] but it is not suitable if the users have different service types. From the point of view of implementation of the admission algorithm in real networks, the availability of the measurements is crucial. Utilization of measurements performed already for other purposes, such as soft handover, might be one alternative because then overhead from additional reporting for the sole purpose of admission control could be avoided. The measurements could be filtered or averaged to avoid the impact of rapid load fluctuations.

The admission control algorithm could optionally estimate the resource consumption of the incoming call (uplink and downlink) [53]. This is a difficult task because the resource requirements of the call might change after the connection establishment if the user moves in the geographical region or increases his or her data transmission activity.

The sum of current load situation and the estimated resource consumption of the new user are compared in the admission algorithm to a threshold value, and the call is blocked or admitted depending on whether the threshold is exceeded or not. At this stage, distinction between handover calls and new calls could be made by reserving a margin for handover calls [54] and thus making it easier for them to get admitted. The threshold that is used in the admission algorithm could be fixed [49] or adjusted dynamically based on instantaneous call dropping rate [49] or unsatisfied users. Dynamic adjustment of the threshold makes the algorithm more complex but also attempts to take into account the time-varying network conditions that affect the objective of the admission control (e.g., dropping probability): A fixed threshold requires that all the factors remain the same in order that it yields the dropping probability.

The admission threshold can also be multivalued. This might be useful to provide some discrimination between users based on, for example, their speed [46]. Another way to use multiple admission thresholds is to provide soft decisions in such a way that in low-load situations calls are always admitted, in high-load situation calls are always blocked, and between those extremes calls are admitted with certain probability [50].

The structure of the network should be taken into account in admission control. If multiple cell layers (micro/macro) are used, the admission control algorithms and thresholds can be different in different layers [54]. This is understandable because one of the reasons for introducing overlapping cell layers is to cater for different needs of users, and therefore it should be reflected in the admission decision as well. Another way to take into account the network structure is to estimate the impact that the admission of a new call would have on adjacent cells [54]. This is relevant because the interference caused by the new call to other cells might drive them to an overload situation. Also, if the admission to one cell is not possible due to its load situation, admission to another cell could be attempted instead.

Prioritization of service types can be attained in the admission control algorithm by providing different requirements for different service types. Low-priority calls might have stricter admission conditions than high-priority calls and they might be required not to affect the quality of high-priority calls [51]. Also, admission conditions for nonreal-time calls can be more flexible because they can be put into a queue due to less stringent delay requirements compared to real-time calls [51].

Even more complicated admission control algorithms than those presented above could be envisaged. For example, soft computing techniques such as fuzzy logic and neural networks could be used in the decision-making [47]. Also, other resource management techniques such as QoS renegotiation could be connected to the admission control and thus provide more flexibility especially in heavily loaded networks.

4.3.4.4 Conclusions

Some relevant aspects of admission control algorithms have been presented in previous studies. The conclusions are summarized here.

Concerning general aspects and performance of admission control, it is stated in [52] that under a heavy load, admission control improves network performance in connection with scheduling algorithms but the magnitude of the improvement is dependent on the scheduling policy used. This implies that the connections between different resource management techniques should be taken into account in the evaluation.

Conclusions about superiority of a certain admission algorithm are difficult to draw. However, some issues could be mentioned based on previous studies. Intelligent admission control based on fuzzy logic and neural networks is said to achieve better performance (higher capacity and lower blocking) than SIR-based admission control [47]. However, clearly the drawback is added complexity. By adding complexity in other ways as well, algorithms with better performance can be obtained. Algorithms using dynamic thresholds provide better performance than fixed-threshold algorithms [49] and interference-based admission control provides better performance than a simple hard blocking algorithm [53]. Further comparison results are provided in [50] where it is shown that more complicated algorithms, modified predictive call admission control, and soft-decision call admission control, yield better performance than simpler algorithms. However, it is also mentioned in [50] that admission control–using measurements might even perform worse than fixed-channel assignment–based admission control due to measurement errors and, in any case, there is no capacity gain if the traffic is not bursty. Thus, the performance of admission algorithms is sometimes highly dependent on the scenario in which they are used.

If some sort of prioritization or differentiation between service types or QoS levels is used in admission control, it is understandable that the performance of those service types that are prioritized is increased. However, in some sources it is shown that even the overall system performance can be enhanced by providing prioritization [51].

Concerning multicell admission control approaches, it is stated in [48] that looking around call admission control appears more flexible than power-based call admission control in nonuniform load situations. Also, in [45] it is mentioned that significant capacity gain can be obtained in nonuniform load situations using multi-cell admission control compared to single-cell admission control. However, reference [55] states that simple single-cell–received power-based admission control performs as well as multicell algorithms or algorithms that attempt to estimate the impact of new calls.

Other aspects of admission control are concluded in [56] where it is mentioned that the GOS for calls can be balanced in the whole cell by using a distance dependent factor in admission control algorithm. In [57], it is said that event-driven measurements yield about the same performance as periodical measurements but they cause less signaling load.

4.3.5   Load Control

The admission control should take care of blocking incoming calls if their admission might jeopardize the quality of ongoing calls. However, in some cases overload situations could arise even if a strict policy is applied at the time of admission. This could be due to, for example, user mobility; users might move to an unfavorable location requiring them to transmit with higher power than initially expected and thus the network load would increase. To overcome this kind of situation, load control is used all the time. Its purpose is to monitor the network load situations based on measurements made in the network, detect possible congestion situations, and apply suitable techniques to resolve the congestion situation.

Many other resource management techniques are related to load control The role of admission control compared to load control was already mentioned above. Load control could utilize the same measurements as a basis for congestion detection as admission control. As a resolution of congestion, load control could trigger forced handovers or affect the power control algorithm by constraining allowed transmission powers. Also, changes in the data rate allocations of users could be carried out.

4.3.5.1   3GPP Constraints/Support

The actual load control algorithms are not standardized, and the main role of standards in connection with load control is to provide support for network load monitoring and procedures to resolve possible congestion situations.

3GPP standards define a measurement model to be used in the network [32], which is composed of measurements in UE/UTRAN, filtering, and measurement reporting (periodic or event-triggered). The actual measurements that should be supported by UE and UTRAN are described as well in [58]. Examples of measurements made by UE are related to common pilot channel (CPICH) receive strength code power (RSCP) and Ec/No (energy per chip per noise density) to be used, for example, with intrafrequency and interfrequency handovers, GSM carrier received signal strength indicator (RSSI) values for intersystem handovers, and UE transmission power. On the other hand, UTRAN must measure received wideband power, transmitted carrier power, and transmitted code power. The received and transmitted total powers in node B are commonly used in connection with load control.

The usability of measurements is further determined by other requirements on the measurement process. Such requirements are reporting range, measurement period, and accuracy (absolute, relative) defined in [33]. For CPICH RSCP measurements, the reporting range is –120...–25 dBm, and the lowest accuracy requirement for intrafrequency measurements is 6 dB. The temporal congestion detection accuracy is affected by the measurement period. Also, whether a certain situation can be designated a congestion or not depends on the accuracy of the measurement.

The fact that UE is required to perform certain measurements does not yet imply that the measurements would be accessible for the centralized load control algorithm located in the network side. This depends on how and when the measurements are delivered or reported to the network. The network can command UE to set up, change, or release measured quantities and affect reporting style (i.e., whether reporting is performed periodically or based on an event-triggered mechanism when certain conditions are fulfilled [25]. The reporting style affects network signaling load and the accuracy of the measurements in the way that it determines how recent values are available in the network side. The measured quantities are divided into six categories: intrafrequency, interfrequency, inter-RAT, traffic volume, quality, UE-internal, and UE positioning measurements. Furthermore, the measured cells are divided for each UE into three mutually exclusive categories (active, monitored, and detected cells). This affects which measurements are available for certain cells.

4.3.5.2   Evaluation Criteria

The objective of load control is to resolve overload situations efficiently. The better the load control algorithm, the less degradation in quality the network experiences when congestion occurs. Thus, the evaluation criterion for the load control efficiency could be anything that measures the ability of the network to deliver the agreed QoS to the calls. Quality degradation could be reflected in increased blocking or dropping rate [59] or reduced throughput [60]. The load control actions can be so diverse and the measurement of the fulfillment of the objectives can vary so that many different criteria can be devised and they could be the same or similar to the criteria used in connection with other resource management techniques (e.g., power control, handover, and admission control). This is also due to the fact that the available load control procedures are at least partially the same as in other resource management technique categories. What load control mainly adds is the framework of congestion measurement and detection and subsequent execution of suitable techniques.

4.3.5.3   Algorithms

Congestion detection in load control requires suitable measurement quantities as a basis. In [61], analogous measures in uplink and downlink directions are defined. In the uplink, noise rise is defined by relating the total interference to the background noise level. In the downlink, the total transmission power is related to the transmission power of some common channel (e.g., pilot channel). In this way, similar load control frameworks in both directions can be attained.

If some quantities measured by UE are used in the load control, they should be reported by UE to the network. Furthermore, the way in which they are reported, in a periodical or event-triggered manner, influences network performance. Too frequent reporting increases the network signaling load. In [62], intrafrequency measurements used mainly for handover are compared with periodical and event-triggered reporting modes. It is shown that periodical reporting can result in better system quality but with the same system quality event-triggered reporting causes a lot less signaling. Similar conclusions are obtained in [63] where the quality of uplink noise rise estimates is compared using periodical and event-triggered reporting.

The stage at which load control actions are triggered, plays an important role. Congestion resolution should be started early enough to prevent futile uncontrolled call dropping and degradation of call quality. In [60] it is pointed out for the uplink direction that the network gets unstable already well below the theoretical capacity. In real networks and situations, many random factors influence the network performance and the propagation environments are very irregular. This has the potential of increasing even further the possibilities for large fluctuations in the network load.

The actions taken in the load control can fall to many different categories. In [64], some examples of possible load control actions are given: deny downlink power-up commands, reduce uplink Eb/No target, reduce throughput of packet traffic, handover to another WCDMA carrier or GSM, decrease bit rates of real-time users, or drop calls in a controlled fashion. The actions mentioned might degrade the call quality of ongoing calls but they should try to avoid situations in which calls get dropped in an uncontrolled manner. In [59] the halving or doubling of the SF of downlink for ongoing calls is proposed as a load control action. Also, in [65] it is mentioned that the data rate of higher data rate users could be downgraded when the node B power exceeds a certain limit. Different procedures for reconfiguring radio resources in load control are mentioned in [60]: limiting the transport format combination set (TFCS) of transport channels (using the transport format combination control message), assignment of a new TFCS (using the transport channel reconfiguration message), and reconfiguring the whole RAB (using the radio access bearer reconfiguration message).

4.3.5.4   Conclusions

Only some general conclusions can be drawn for this group of diverse algorithms. What is usually common for a load control framework is the usage of some sort of thresholds for detecting congestion. In [60] it is pointed out that the selection of optimal thresholds for the load control depends on the trade-off between efficient resource utilization and rapid congestion detection. Load control actions should not be started too early in order not to sacrifice the network capacity but, on the other hand, taking actions too late can result in high degradation of call quality or even call dropping. The efficiency of the load control depends on the scenario. The results in [59] indicate that for the downlink operation, the gains of the presented load control actions are influenced by whether the scenario is power-limited or code-limited. Concerning different service types in load control, in [60] it is brought up that services with strict delay requirements can be protected by reducing data rates of delay-tolerant service types (best effort services).

4.3.6 Scheduling

The purpose of scheduling is to control the transmissions of delay tolerant services types (i.e., nonreal-time traffic). Scheduling works on a relatively fine time scale attempting to utilize efficiently the capacity remaining after all real-time users have been served. The idea is to decide who is allowed to transmit, with what rate, when, and for how long. Different scheduling algorithms can be used depending on how much weight is given to specific objectives of scheduling—for example, fairness and total throughput.

4.3.6.1 3GPP Constraints/Support

One of the responsibilities of the MAC layer in the UE side is to take care of priority handling between multiple data streams [66]. RRC can assign a priority between 1 and 8 to each logical channel and the transport format combination (TFC) selection will be performed by taking into account the assigned priorities. Furthermore, the TFCs will be chosen based on the TFCSs received from the network. Some additional requirements for TFC selection are given in [66].

An issue also affecting scheduling is the standardization of HSDPA. The channel used as a basis for HSDPA is HS-PDSCH. The SF used in HS-PDSCH is always 16. In HSDPA, it would be possible to perform fast scheduling in node B. However, even with the standardization of HSDPA and HS-PDSCH, the actual scheduling algorithm would still not be standardized—just the procedures enabling scheduling. In [67], system simulation recommendations related to the evaluation of features of HSDPA are presented. Concerning packet scheduling, different scheduling periods (1.5 and 15 slots) and algorithms (round robin and C/I) are proposed for evaluation.

4.3.6.2 Evaluation Criteria

Scheduling controls the resources assigned to nonreal-time users. Nonreal-time service types tolerate delay, and this is what is exploited in scheduling algorithms. Thus, delay measured in different ways should be used as a criterion in the assessment of scheduling algorithms. In different forms, it has been taken into account [e.g., in [68] (SPDU delay), in [69] (nominal delay) and in [70] (delay)]. A metric similar to delay is the mean time in the queue (i.e., the time user has to wait before being able to transmit [71]). Furthermore, variations in the experienced delay (i.e., delay jitter can be used as a metric for scheduling algorithm evaluation).

The utilization of real-time service transmitting constantly can be measured in many cases simply using the number of calls. However, for bursty nonreal-time users, a very relevant performance metric the throughput that indicates how much data is sent per unit of time. Scheduling algorithms should attempt to maximize the throughput and thus take full advantage of the resources of the network. Throughput can be measured (e.g., per user or per cell). Similarly, the achieved bit rate can also be expressed for a certain class of users [72].

More information than just the mean or aggregated bit rate is provided by the distribution of the allocated or used bit rates. It can be expressed as a cumulative distribution function or discrete probability distribution function. The distribution indicates, for example, shifts in the bit rate allocations as the network load or scheduling algorithm changes.

Quality metrics used in connection with scheduling algorithms are blocking and dropping probabilities [73], outage probability [72], and packet loss probability or rate [74]. The last metric is also related to retransmission strategy and depends on it.

The total available transmission power of the base station is always a limited and scarce resource. Thus, scheduling policies could also consider this as one metric indicating network load and thus capacity utilization [68]. A similar metric in the uplink direction is some quantity derived from interference (e.g., noise rise distribution [65]). Both of these two quantities can be measured in the network side.

Finally, the energy required to correctly transmit one bit takes into account different channel conditions of users [69]. It is desirable to have a minimal amount of energy (or power) to maximize the throughput, but unfortunately this can easily lead to unfairness among users: Users in favorable channel conditions are served more than users in unfavorable conditions.

4.3.6.3   Algorithms

There are plenty of different scheduling algorithms that have been investigated in the literature. On a general level, the scheduling strategy can be either code division–based, so that many users are allocated low bit rates at the same time, or time division–based, so that only a few users are allowed to transmit at the same time but with high bit rates. These two cases have been investigated and are mentioned (e.g., in [75]). Many different issues affect the selection of the scheduling algorithm; it depends on which of the evaluation criteria presented before are emphasized. Furthermore, such issues as traffic and physical layer characteristics have an effect on the optimality of certain scheduling policy.

Algorithms related to scheduling of different data streams in one UE select the TFC for each frame based on the TFCS received from the network. Priorities to different data streams (e.g., voice, signaling, and WWW) can be assigned and utilized in the scheduling algorithm.

Scheduling the transmissions of different users either in the uplink or downlink direction is another issue compared to the scheduling of different data streams in one UE. Restricting the uplink allowed TFCs by sending from the network side appropriate TFCS has been investigated in [65].

In the scheduling algorithms, some measurement is required as a basis for estimating the instantaneous capacity utilization of the network. The TFCs or processing gains of the users are then chosen based on remaining resource (e.g., after all real-time users have been allocated powers). In the uplink, as an estimate of the bandwidth utilization the product of data rate and SIR could be used [70]. A common network load measurement used in other RRM algorithms as well, is the uplink received power or interference, which can be measured in the network side and applied as a criterion for scheduling decisions [72].

In round robin techniques, users are cyclically allowed to transmit irrespective of their channel conditions. Similar to the round robin technique, in fastest first users are cycled through but the user to transmit first is chosen based on who can finish the transmission first [75]. Round robin is not so efficient in terms of throughput because the channel conditions are not taken into account and thus users in an unfavorable situation might also be allowed to transmit. Contrary to that, C/I-based scheduling policies exploit the knowledge of the channel conditions and attempt to allocate bandwidth to users with favorable channel conditions. This could be achieved by delaying transmission by a time that depends on the C/I value [71]. The problem with C/I compared to round robin is the lack of fairness: Certain users get far more resources than others. As a compromise between round robin and C/I-based scheduling strategies, the modified round robin technique is proposed in [69]. The idea is to avoid transmitting during deep fades that are determined by comparing the instantaneous C/I value to the average calculated from past. In a similar approach, proportional fair scheduling is compared to round robin and C/I algorithms in [76]. Also, fair resource scheduling in [68] can be seen as a proposal attempting to increase fairness compared to C/I but still retaining the channel conditions as one criterion for the scheduling decision.

Two algorithms based on the usage of queue size as scheduling criterion, either by giving priority to shortest or longest queue, are presented in [52]. A slightly similar technique is presented in [74] where the concept of service credit is introduced to measure the deviation between the bit rate requested by the user and the bit rate offered to the user.

Multiple users can be allowed to transmit at the same time (e.g., by sharing the resources (power) among all users equally (equal resource sharing) [72]. Similarly, in the equal weight scheme, the channel conditions are taken into account when deciding data rates to allocate simultaneously to multiple users, and the effective system bandwidth is shared equally between users. On the other hand, in the equal rate scheme the same data rate is given to each user at the same time irrespective of channel conditions [75].

With the introduction of the HSDPA concept, the actual scheduling algorithm still remains unspecified. Thus, it is possible to design and implement arbitrary scheduling policies in a similar way without HSDPA. Some examples for HSDPA scheduling algorithms (fair throughput, fair resource, C/I, proportional fair throughput, proportional fair resource, and maximum C/I) are presented in [73]. Furthermore, they are divided into two categories based on the scheduling interval: slow (20...100 ms) and fast [per transmit time interval (TTI)].

4.3.6.4   Conclusions

The comparison of the performance of different scheduling algorithms is difficult because it depends on what is considered important. Also, the evaluation environment (e.g., service types and QoS levels) affects the results, and thus even the conclusions presented below might depend to some extent at least on the conditions in which the evaluation was carried out. If the objective is to maximize throughput of the system, the channel conditions should be taken into account, and users with favorable conditions should be allowed to transmit. However, in that case fairness among users is violated. Better throughput than with round robin can be attained with, for example, modified round robin [69], but it is still worse than throughput attained using C/I-based scheduling. However, for C/I the fairness is lower and it is sensitive to network conditions [69]. Also, fairness and differentiation between QoS classes can be achieved with fair resource scheduling [68] compared to round robin. For differentiation between service classes, the algorithm presented in [70] can be used.

In [72] it is mentioned that highest throughputs can be attained by allowing only a few users to transmit at a time. Prioritization of the shortest queue compared to prioritization of the longest queue yields better performance [52]. The impact of the load situation is mentioned in [65] where it is pointed out that in high-load situations stability in the interference situation is brought by the fact that there is less variation in the selected TFCs. In the same source, it is also concluded that the scheduling period has an impact on the results when there is a lot of variation in the offered load. Concerning the usage of instantaneous C/I values versus average C/I values in the scheduling policy, in [65] it is indicated that instantaneous values provide better performance in terms of outage. In general, better performance in terms of throughput and delay can be attained by reserving the whole bandwidth to one user at a time [75].

4.3.7   Miscellaneous Techniques

Techniques presented in the previous chapters are the most essential for the operation of an UMTS network. There are, however, many more, which are not necessary, but can improve performance of the network. Some examples of such techniques are presented in this chapter.

4.3.7.1   Preemption and Dropping Policies

Preemption is based on the prioritization of calls. The technique is simple: If a call of higher priority cannot be admitted to a network due to high load, another call, of lower priority is dropped to free resources. The same may happen if an ongoing high-priority connection is at risk of outage, or a congestion is about to start: A preventive dropping may be executed on calls of lower priority [46]. This is especially important in emergency situations, when despite a high load, a certain call must be admitted and offered acceptable quality. The preemption policy is not standardized and therefore is a subject of research [77]. The procedure may end with degradation of the lower priority call, instead of dropping.

UMTS has, however, some tools that can be utilized for preemption algorithms. The first is the eMLPP mechanism inherited from GSM. It makes it possible to define up to seven priority classes, which are assigned to certain users either statically, or dynamically. This framework was a base for the implementation of emergency calls treatment. The other is a new feature introduced in UMTS: QoS. One of the QoS parameters (at the highest level, UMTS bearer) is "allocation and retention priority" (range: 1–3). This parameter, together with others (e.g., service class or traffic handling priority) may be mapped through lower planes onto access stratum and RAB priority (the mapping method is not specified). There are 16 levels available and preemption vulnerability and preemption capability flags may be used. There is also a clearly defined method to deal with them. When an RNC receives from the network a RAB establishment request it must follow given rules [45]:

1. The values of the last received *preemption vulnerability* and *priority level* shall prevail.
2. If the *preemption capability* is set to "may trigger pre-emption," then this allocation request may trigger the preemption procedure.
3. If the *preemption capability* is set to "shall not trigger preemption," then this allocation request shall not trigger the preemption procedure.
4. If the *preemption vulnerability* is set to "preemptable," then this connection shall be included in the preemption process.
5. If the *preemption vulnerability* is set to "not preemptable," then this connection shall not be included in the preemption process.
6. If the *priority level* is set to "no priority used" the given values for the *preemption capability* and *preemption vulnerability* shall not be considered. Instead the values "shall not trigger pre-emption" and "not preemptable" shall prevail.

The actual decision does not always have to lead to call drop. Other policies may have a similar effect but cause less frustration to users. The first possibility is renegotiating of QoS of the lower priority call, if its degradation can free enough resources. This is especially relevant with NRT services, which can stand much longer delays than RT calls. Therefore a data session can be rather suspended than dropped, if the congestion is temporary.

4.3.7.2 QoS Techniques

QoS was introduced in GPRS, but it is going to be fully exploited only in UMTS. It enables offering a service with different quality levels (selected either according to subscription or network conditions). This brings about two new possibilities: QoS negotiation and renegotiation.

QoS negotiation is the process at the beginning of a call: A user requests certain quality, but if it is not available due to network load, he or she is offered a lower quality. This process may be invisible to the user and performed between the core network, which is responsible for the complete connection and the RNC, which controls radio resources (Section 4.3.4.1). An example of this may be selecting the AMR level for a speech call.

QoS renegotiation happens when there is a need to take some resources from an ongoing call (e.g., due to worse radio conditions or a preemption procedure). The initial quality may be later resumed. Similarly, a call established with a lower quality level, which was the result of QoS negotiation may be reconfigured (its QoS renegotiated) to the level requested initially. The renegotiation, like the negotiation, may be invisible to the user.

Neither of the procedures is standardized in ETSI documents. There are, however, available tools to implement such techniques. The first is the QoS architecture, then there is the admission process (Section 4.3.4.1) and tools enabling reconfiguration of an ongoing call [25].

An example algorithm may take into account users' priorities. QoS traffic classes are assigned priorities, and each of them is treated differently in case of congestion: RT calls of low priorities are dropped; NRT sessions are delayed; and when this does not release enough resources QoS negotiation is used. As a result, lower blocking probability is achieved in the system [78].

The change of QoS parameters is to enable the network to reconfigure physical resources. Beside obvious solutions consisting of decreasing quality of a physical channel, there are also more sophisticated methods. One of them is dynamic switching between dedicated and shared channels [79]. In dedicated channels the control part is transmitted permanently, whereas in shared channels it is activated only when data is transmitted. This leads to a conclusion that pushing some part of the traffic to a shared domain may decrease the load.

4.3.7.3   Multiple Carriers and Scrambling Code Assignment

One cell may use more then one OVSF code tree, but in this case there is a need to assign more than one scrambling code to the cell. This technique makes it possible to multiply downlink capacity, but poses an additional challenge: how to distribute users among available scrambling codes. The best results (in terms of downlink interference) are achieved when one scrambling code is fully occupied [80]. This is because within a scrambling code interference is minimized with the orthogonality. The other scrambling code should be is assigned to users located near the base station, where interference is low because of good propagation conditions.

4.3.7.4   Node B Clustering

In real networks cells may be loaded unevenly. As long as there re enough resources, it is not a problem; however at some moment more loaded nodes-B may reach the upper limit of their transmission power. At the same time those less loaded have still enough spare resources. To prevent such a situation, they may be clustered, and each cluster shares resources (available transmission power) [81].

If an uneven load is caused by an unusual situation it may form a hotspot. A hotspot located in one cell is likely to cause a congestion—resources in this one cell will not be enough to serve the traffic. Distributing load with a handover amendment may lead to interference problems. A solution to this could be adaptive cell sectorization—a site is able to rotate sectors in such way that the hotspot is covered by two cells. Simulations suggest that the benefit of this technique is that the bigger, more demanding services were used in the hotspot [82].

4.3.8   TDD-Mode RMTs

TDD is the other access technology specified for UMTS, besides the more popular FDD. TDD uses the same frequency band for transmission in both directions. It enables efficient resource utilization for unsymmetrical services and therefore is recommended for hotspots, where high-quality data services form a significant part of the total traffic.

4.3.8.1 3GPP Constraints/Support

In TDD there is 15-slot time frame, but the slots may be allocated either for the uplink, or for the downlink. The minimum allocation is 2 TS in the downlink and 13 in the uplink, or 14 TS in the downlink and 1 in the uplink. In each time slot several codes may be used if radio conditions allow for that (OVSF codes are used with the spreading factor from 1 to 16; in the downlink SF is always 16, in the uplink it may vary). The combination of time slots and codes allows very flexible and efficient resource allocation.

Power control is very different from the one standardized for FDD. The main difference is due to discontinuous transmission, which constrains usage of the inner power loop (no continuous feedback). Therefore in the uplink only the open power loop is used: At the beginning of each allocated time slot a UE estimates required power. In the downlink there is an inner loop power control, but the frequency of the TPC commands depends on timeslot allocation in the uplink (in any case it does not exceed 800 Hz).

The usage of the same frequency band for both directions requires very precise synchronization between both ends of a connection, but also between cells to avoid overlapping of time slots and unnecessary interference. The slot allocation, which is different in each cell, is also the reason for handover implementation: In TDD only hard handover is possible. A UE that is about to cross a cell boundary must reschedule the transmission and reception according to the resource allocation in the new cell. Therefore the connection to the old cell must be released.

4.3.8.2 TDD Problems and RMTs

Implementation of TDD may encounter problems specific to this technology. First is the interference between downlink and uplink. Furthermore, the proximity of the FDD uplink band may be a source of additional interference. The problem of power management is more difficult in TDD as compared to FDD, for the former has less efficient power control. Research shows that the impact of FDD is negligible. In the other direction, the impact of TDD onto FDD/UL is negligible also if base stations of the technologies are separated by at least 200m. The interference among UEs is, however, stronger and requires careful consideration at the network planning phase. The next problem specific to TDD is synchronization of the network. Finally, TDD coverage is much smaller than in case of FDD. This is because in discontinuous transmission mean power is lower, which is reflected in cell range. Because of that, and the interference issues mentioned above, TDD is regarded mainly as an indoor technology.

The most interesting issue is resource allocation, which is not standardized. In TDD there are two types of channel allocation: slow, which allocates resource among cells, and fast, which is performed within each cell and enables it to distribute available resources among connections. There has not yet been much research done in this area, but it has been shown that homogenous time slot allocation in the whole network results in better average performance [83]. A dynamic resource allocation, which prioritizes users with the best connection quality, offers better performance than a random fixed allocation, which makes it possible only to distribute interference uniformly in the network [84].

The power control mechanism of TDD could be enhanced, if instead of simple "up" or "down" commands a difference between SIR and SIR_{target} is sent. According to that information the PC step could be calculated. The dynamic step could compensate slow power control [85]. Admission control is related to resource allocation. In TDD the load (expressed with load control procedures) is not always the best indicator—two timeslots may exercise different outage probabilities despite the fact that their loads are very similar. Therefore a formula to estimate the outage probability has been proposed and an admission control algorithm, which aims at minimizing the probability of ongoing calls and those that are being admitted [86]. The handover of TDD can be based on a relative threshold algorithm, similarly to FDD soft handover. To enable a quick reaction a candidate set should be selected from all possible cells. The selection could be based on the pilot channel received power so that all selected cells would offer good enough quality for the connection. If one of the candidate cells becomes better by a certain threshold than the current cell, the handover is performed [87].

4.3.9   UMTS-GSM/GPRS RMTs

The UMTS network is expected to be deployed, especially initially, mainly in regions where the population density is high and where demand for high data rate services is expected to exist. The existing GSM/GPRS networks will coexist with UMTS and provide continuous coverage also in regions that are not covered by UMTS. Furthermore, the core network of UMTS in Release 99 is the same as for GSM/GPRS. In subsequent releases (Release 4 and Release 5), the core network undergoes some modifications.

The coexistence of two radio access technologies will pose requirements on inter-RAT radio resource management. The basis for resource management in UMTS is provided by measurements made by UE and UTRAN. Similarly, inter-RAT measurements are defined in 3GPP specifications to provide a basis for inter-RAT resource management [25]. Examples of measurements that could be used to control traffic between GSM and UMTS are CPICH *Ec/No*, CPICH RSCP, and GSM carrier RSSI [25]. Furthermore, some reporting events are defined for aiding in triggering RRM actions. An example is the reporting event 3a: "The estimated quality of the currently used UTRAN frequency is below a certain threshold and the estimated quality of the other system is above a certain threshold." Other inter-RAT reporting events (3b–3d) concern the quality of the other system (GSM) and changes in the quality. Whether CPICH *Ec/No* or RSCP should be used in inter-RAT RRM has been investigated in [88]. It is concluded there that the RSCP value gives a better indication of the UMTS coverage but the required accuracy for that measurement is lower than for *Ec/No*.

One issue that is relevant from the point of view of performing inter-RAT measurements is the compressed mode. Normal transmission has to be interrupted and a transmission gap created to be able to make measurements from GSM cells. This affects power control and deteriorates system performance. Thus, the usage of the compressed mode should be optimized [88].

After the appropriate measurements from UMTS and GSM networks are available, suitable procedures for performing inter-RAT RRM are required. Some tools for that purpose are described in [25]. These include inter-RAT handover from/to UTRAN, inter-RAT cell reselection from/to UTRAN, and inter-RAT cell change from/to UTRAN. Furthermore, some requirements (e.g., for inter-RAT handover) are given in [33].

Some results concerning inter-RAT handover between UMTS and GSM are presented in [88]. In [89], the coexistence of multiple cell layers and radio access technologies are investigated. Procedures used for controlling resources are intersystem handover, directed retry, and intersystem cell reselection. Both real-time and nonreal-time service types are taken into account in the analysis. The idea is to provide a resource management scheme that takes into account all the network types and utilizes the knowledge about individual cell load situations. The capacity gain attained using common resource management is shown to depend on the service type and it is usually very relevant.

4.3.10 Summary and Guidelines for Implementation

This chapter covered a wide range of resource management techniques applicable to UMTS networks. The 3GPP standards, algorithms, evaluation criteria, and conclusions were addressed for each class of technique. Concerning practical implementation of the techniques in real networks, it should be kept in mind that some conclusions presented might be valid only in certain scenarios or under the assumptions made in the studies. Also, many results had been obtained in the studies using simulators and very regular scenarios and thus many unaccounted factors might influence the actual gains of the techniques in real situations. However, a good basis is provided in the described techniques and a starting point given for real implementation.

In the evaluation of the RMTs in real trial networks, the same evaluation criteria would still be applicable (e.g., dropping/blocking/outage/packet loss probability, number of HO operations, soft HO overhead, TX power, noise rise, delay, delay jitter, throughput or bit rate, and energy per bit. Some evaluation criteria could be calculated per user, cell, node B, network, service type, QoS level or priority class. Also, the impact of RMT could be measured in an absolute value or as a gain relative to some reference case. Furthermore, arbitrary combinations of the evaluation criteria could be used (e.g., $GOS = 10 \ P_{drop} + P_{block}$). However, in real situations, more emphasis should be placed on the revenues that the operator can gain from the usage of the technique because this is how the positive impact should ultimately manifest itself.

The resources to be controlled (e.g., OVSF codes, power, and links) should be clearly identified, and suitable techniques for affecting each of the resources should be adopted. The advantages and drawbacks (e.g., soft handover gain in downlink) of each technique has to be analyzed. Also, the parameters for each technique to be adjusted (e.g., soft handover add and drop thresholds) should be identified and practical values for them established. Preliminary indications on the trends of network performance as a function of the RMT parameters can be obtained using a simulation approach. Measurements made by UE/UTRAN and the reporting of UE measurements to UTRAN should be taken into account. The procedures and tools available for RRM, delays in executing them, and possible imperfections (e.g., misinterpretations of power control commands) should be identified. Here, the discussion concerning different groups of RMTs has been done separately but in reality the interactions between different RMTs should be considered as well.

References

[1]   UMTS Forum, "Minimum Spectrum Demand Per Public Terrestrial UMTS Operator in the Initial Phase," 1998.

[2]   3GPP, TSG Radio Access Networks, "BS Radio transmission and Reception (FDD)," 3GPP TS 25.104 (V.4.7.0), March 2003.

[3]   3GPP, TSG Services and System Aspects, "General UMTS Architecture," 3GPP TS 23.101 (V.4.0.0), April 2001.

[4]   3GPP, TSG Radio Access Network, "UTRAN Overall Description," 3GPP TS 25.401 (V.4.6.0), December 2002.

[5]   3GPP, TSG Radio Access Network, "UE Radio Transmission and Reception (TDD)," 3GPP TS 25.102 (V.4.7.0), December 2002.

[6]   3GPP, TSG Radio Access Network, "UTRAN Iu Interface: General Aspects and Principles," 3GPP TS 25.410 (V4.5.0), September 2002.

[7]   3GPP, TSG Radio Access Network, "UTRAN Iu interface RANAP Signaling," 3GPP TS 25.413 (V4.9.0), June 2003.

[8]   3GPP, TSG Radio Access Network, "UTRAN Iur Interface General Aspects and Principles," 3GPP TS 25.420 (V4.2.0), March 2002.

[9]   3GPP, TSG Radio Access Network, "UTRAN Iur Interface RNSAP Signaling," 3GPP TS 25.423, (V4.9.0), June 2003.

[10] 3GPP, TSG Radio Access Network, "UTRAN Iub Interface: General Aspects and Principles," 3GPP TS 25.430 (V4.4.0), September 2002.

[11] 3GPP, TSG Radio Access Network, "UTRAN Iub Interface NBAP Signaling," 3GPP TS 25.433 (V4.9.0), June 2003.

[12] 3GPP, TSG Radio Access Network, "Radio Interface Protocol Architecture," 3GPP TS 25.301 (V4.4.0), September 2002.

[13] 3GPP, TSG Radio Access Network, "Medium Access Control (MAC) Protocol Specification," 3GPP TS 25.321 (V4.8.0), March 2003.

[14] 3GPP, TSG Radio Access Network, "Radio Link Control (RLC) Protocol Specification," 3GPP TS 25.322 (V4.9.0), June 2003.

[15] 3GPP, TSG Radio Access Network, "Packet Data Convergence Protocol (PDCP) Specification," 3GPP TS 25.323 (V4.6.0), September 2002.

[16] 3GPP, TSG Radio Access Network, "Broadcast/Multicast Control (BMC)," 3GPP TS 25.324 (V4.3.0), March 2003.

[17] UMTS Forum, "The Path Toward UMTS Technologies for the Information Society," 1998.

[18] 3GPP, TSG Services and System Aspects, "UMTS Phase 1 Release 99," 3G TS 22.100 (V3.7.0), October 2001.

[19] http://www.3gpp.org/specs/releases-contents.htm.

[20] Holma, H., A. Toskala, *WCDMA for UMTS, Radio Access for 3rd Generation Mobile Communications,* New York: John Wiley & Sons, 2001.

[21] Economou, P., and A. Grillakis, "Examining COSMOTE's UMTS Network Assumptions and Planning Approaches Utilizing Existing GSM Network Resources and Data," *IIR Conference*, Prague 2002.

[22] Laiho, J., A. Wacker, T. Novosad, "Radio Network Planning and Optimization for UMTS," New York: JohnWiley & Sons, 2001.

[23] 3GPP, TSG Radio Access Network, "Physical Layer Procedures (FDD)," 3GPP TS 25.214, (V 4.6.0), March 2003.

[24] 3GPP, TSG Radio Access Network, "User Equipment (UE) Radio Transmission and Reception (FDD)," 3GPP TS 25.101, (V 4.8.0), June 2003.

[25] 3GPP, TSG Radio Access Network, "Radio Resource Control (RRC): Protocol Specification," 3GPP TS 25.331, (V 5.5.0), June 2003.

[26] Staehle, D., K. Leibnitz, and K. Heck, "Effects of Soft Handover on the UMTS Downlink Performance," *Proceedings of 56th VTC*, Vancouver, British Columbia, Canada, September 24–29, 2002, pp. 960–964.

[27] Akhtar, S., S. A. Malik, and D. Zeghlache, "A Comparative Study of Power Control Strategies for Soft Handover in UTRA FDD WCDMA System," *Proceedings of 53rd VTC*, Rhodes, Greece, May 6–9, 2001, pp. 2680–2684.

[28] Schinnenburg, M., I. Forkel, and B. Haverkamp, "Realization and Optimization of Soft and Softer Handover in UMTS Networks," *Proceedings of European Personal and Mobile Communications Conference*, Glasgow, Scotland, U.K., April 2003.

[29] Hamabe, K., "Adjustment Loop Transmit Power Control During Soft Handover in CDMA Cellular Systems," *Proceedings of 52nd VTC*, Boston, MA, September 24–28, 2000, pp. 1519–1523.

[30] Nuaymi1, L., X. Lagrange1, and P. Godlewski, "A Power Control Algorithm for 3G WCDMA System," *Proceedings of European Wireless Conference*, Florence, Italy, February 25–28, 2002.

[31] Takano, N., and K. Hamabe, "Enhancement of Site Selection Diversity Transmit Power Control in CDMA Cellular Systems," *Proceedings of 54th VTC*, Atlantic City, NJ, October 7–11, 2001, pp. 635–639.

[32] 3GPP, TSG Radio Access Network, "Services Provided by the Physical Layer," 3GPP TS 25.302 (V 5.5.0), June 2003.

[33] 3GPP, TSG Radio Access Network, "Requirements for Support of Radio Resource Management (FDD)," 3GPP TS 25.133 (V 5.7.0), July 2003.

[34] Moghaddam, S. S., V. T. Vakili, and A. Falahati, "New Handoff Initiation Algorithm," *Proceedings of 52nd VTC*, Boston, MA, September 24–28, 2000, pp. 1567–1574.

[35] Jabbari, B., and W. F. Fuhrmann, "Teletraffic Modeling and Analysis of Flexible Hierarchical Cellular Networks with Speed-Sensitive Handoff Strategy," *IEEE Journal on Selected Areas in Communications*, Vol. 15, No. 8, October 1997, pp. 1539–1548.

[36] Chen, Y., and L. G. Cuthbert, "Downlink Performance of Different Soft Handover Algorithms in 3G Multiservice Environments," *Proceedings of the IEEE 4th International Conference on Mobile and Wireless Communications Network (MWCN 2002)*, Stockholm, Sweden, September 9–11, 2002, pp. 406–410.

[37] Sipila, K., et al., "Soft Handover Gains in a Fast Power Controlled WCDMA Uplink," *Proceedings of 49th VTC*, Houston, TX, May 16–20, 1999, pp. 1594–1598.

[38] Flanagan, J. A., and T. Novosad, "WCDMA Network Cost Function Minimization for Soft Handover Optimization with Variable User Load," *Proceedings of 56th VTC*, Vancouver, British Columbia, Canada, September 24–29, 2002, pp. 2224–2228.

[39] Yang, X., S. Ghaheri-Niri, and R. Tafazolli, "Enhanced Soft Handover Algorithms for UMTS System," *Proceedings of 52nd VTC*, Boston, MA, September 24–28, 2000, pp. 1539–1543.

[40]   Laiho-Steffens, J., et al., "Comparison of Three Diversity Handover Algorithms by Using Measured Propagation Data," *Proceedings of 49th VTC*, Houston, TX, May 16–20, 1999, pp. 1370–1374.

[41]   3GPP, TSG Radio Access Network, "Radio Resource Management Strategies," 3GPP TR 25.922 (V 4.2.0), March 2002.

[42]   Hwang, S.-H., S.-L. Kim, "Soft Handoff Algorithm with Variable Thresholds in CDMA Cellular Systems," *IEE Electronics Letters*, Vol. 33 (19), 1997, pp. 1602–1603.

[43]   Valkealahti, K., et al., "WCDMA Common Pilot Power Control for Load and Coverage Balancing," *Proceedings of the 13th IEEE International Symposium on Personal, Indoor and Mobile Radio Communications*, Lisbon, Portugal, September 15–18, 2002, pp. 1412–1416.

[44]   Castro, J. P., *The UMTS Network and Radio Access Technology*, New York: John Wiley & Sons, 2001.

[45]   3GPP, TSG Radio Access Network, "UTRAN Iu Interface RANAP Signaling," 3GPP TS 25.413 (V5.5.0), June 2003.

[46]   Badia, L., M. Zorzi, and A. Gazzini, "On the Impact of User Mobility on Call Admission Control in WCDMA Systems," *Proceedings of 56th VTC*, Vancouver, British Columbia, Canada, September 24–29, 2002, pp. 121–126.

[47]   Chang, C.-J., et al., "Intelligent Call Admission Control for Differentiated QoS Provisioning in Wideband CDMA Cellular Systems," *Proceedings of 52nd VTC*, Boston, MA, September 24–28, 2000, pp. 1057–1063.

[48]   Capone, A., and S. Redana, "Call Admission Control Techniques for UMTS," *Proceedings of 54th VTC*, Atlantic City, NJ, October 7–11, 2001, pp. 925–929.

[49]   Dimitriou, N., G. Sfikas, and R. Tafazolli, "Call Admission Policies for UMTS," *Proceedings of 51st VTC*, Tokyo, Japan, May 15–18, 2000, pp. 1420–1424.

[50]   Phan-Van, V., and D. D. Luong, "Capacity Enhancement with Simple and Robust Soft-Decision Call Admission Control for WCDMA Mobile Cellular PCNs," *Proceedings of 54th Vtc*, Atlantic City, NJ, October 7–11, 2001, pp. 1349–1353.

[51]   Akhtar, S., S. A. Malik, and D. Zeghlache, "Prioritized Admission Control for Mixed Services in UMTS WCDMA Networks," *Proceedings of the 13th IEEE International Symposium on Personal, Indoor and Mobile Radio Communications*, San Diego, CA, September 30–October 3, 2001, pp. 133–137.

[52]   Kazmi, M., P. Godlewski, and C. Cordier, "Admission Control Strategy and Scheduling Algorithms for Downlink Packet Transmission in WCDMA," *Proceedings of 52nd VTC*, Boston, MA, September 24–28, 2000, pp. 674–680.

[53]   Holma, H., and J. Laakso, "Uplink Admission Control and Soft Capacity with MUD in CDMA," *Proceedings of 50th VTC*, Amsterdam, Netherlands, September 19–22, 1999, pp. 431–435.

[54]   Ying, W., et al., "Call Admission Control in Hierarchical Cell Structure," *Proceedings of 55th VTC*, Birmingham, Alabama, May 6–9, 2002, pp. 1955–1959.

[55]   Redana, S., and A. Capone, "Received Power-Based Call Admission Control Techniques for UMTS Uplink," *Proceedings of 56th VTC*, Vancouver, British Columbia, Canada, September 24–29, 2002, pp. 2206–2210.

[56]   Dimitriou, N., and R. Tafazolli, "Resource Management Issues for UMTS," *Proceedings of First International Conference on 3G Mobile Communication Technologies*, London, May 27–29, 2000, pp. 401–405.

[57]   Gunnarsson, F., et al., "Uplink Admission Control in WCDMA Based on Relative Load Estimates," *Proceedings of the IEEE International Conference on Communications*, New York, April 28–May 2, 2002, pp. 3091–3095.

[58]   3GPP, TSG Radio Access Network, "Physical layer—Measurements (FDD)," 3GPP TS 25.215 (V 4.7.0), June 2003.

[59] Rave, W., et al., "Evaluation of Load Control Strategies in an UTRA/FDD Network," *Proceedings of 53rd VTC*, Rhodes, Greece, May 6–9, 2001, pp. 2710–2714.

[60] Sachs, J., T. Balon, and M. Meyer, "Congestion Control in WCDMA with Respect to Different Service Classes," Proceedings of European Wireless Conference, Munich, Germany, October 7–8, 1999.

[61] Muckenheim, J., and U. Bernhard, "A Framework for Load Control in 3G CDMA Networks," *Proceedings of the IEEE Global Telecommunications Conference*, San Antonio, TX, November 25–29, 2001, pp. 3738–3742.

[62] Hiltunen, K., N. Binucci, and J. Bergstrom, "Comparison Between the Periodic and Event-Triggered Intrafrequency Handover Measurement Reporting in WCDMA," *Proceedings of the IEEE Wireless Communications and Networking Conference*, Chicago, September 23–28, 2000, pp. 471–475.

[63] Lundin, E. G., F. Gunnarsson, and F. Gustafsson, "Uplink Load Estimates in WCDMA with Different Availability of Measurements," *Proceedings of 57th VTC*, Jeju, Korea, April 22–25, 2003.

[64] Laiho, J., A. Wacker, and T. Novosad, *Radio Network Planning and Optimization for UMTS*, New York: John Wiley & Sons, 2002.

[65] Dimou, K., et al., "Performance of Uplink Packet Services in WCDMA," *Proceedings of 57th VTC*, Jeju, Korea, April. 22–25, 2003.

[66] 3GPP, TSG Radio Access Network, "Medium Access Control (MAC) Protocol Specification," 3GPP TS 25.321 (V5.5.0), June 2003.

[67] 3GPP, TSG Radio Access Network, "Physical Layer Aspects of UTRA High Speed Downlink Packet Access," 3GPP TR 25.848, (V 4.0.0), April 2001.

[68] Malik, S. A., and D. Zeghlache, "Improving Throughput and Fairness on the Downlink Shared Channel in UMTS WCDMA Networks," *Proceedings of European Wireless Conference*, Florence, Italy, February 25–28, 2002.

[69] López, I., et al., "Downlink Radio Resource Management for IP Packet Services in UMTS," *Proceedings of 53rd VTC*, Rhodes, Greece, May 6–9, 2001, pp. 2387–2391.

[70] Tian, D., and J. Zhu, "A QoS-Oriented Bandwidth Scheduling Scheme on 3G WCDMA Air Interface," *Proceedings of International Conferences on Info-tech and Info-net*, Beijing, China, October 29–November 1, 2001, pp. 139–144.

[71] Carciofi, C., and P. Grazioso, "Radio Resource Management Strategies for Packet Data Services In UMTS," *Proceedings of 53rd VTC*, Rhodes, Greece, May 6–9, 2001, pp. 1012–1016.

[72] Abrardo, A., et al., "Performance Analysis of a Packet Scheduling Policy for a DS-CDMA Cellular System," *Proceedings of 53rd VTC*, Rhodes, Greece, May 6–9, 2001, pp. 2214–2218.

[73] Frederiksen, F., T. E. Kolding, and P.E. Mogensen, "Performance Aspects of WCDMA Systems with High-Speed Downlink Packet Access (HSDPA)," *Proceedings of 56th VTC*, Vancouver, British Columbia, Canada, September 24–29, 2002, pp. 477–481.

[74] Almajano, L., and J. Pérez-Romero, "Packet Scheduling Algorithms for Interactive and Streaming Service Under QoS Guarantee in a CDMA System," *Proceedings of 56th VTC*, Vancouver, British Columbia, Canada, September 24–29, 2002, pp. 1651-1661.

[75] Vaanithamby, R., and E. S. Sousa, "Resource Allocation and Scheduling Schemes for WCDMA Downlinks," *Proceedings of IEEE International Conference on Communications*, St. Petersburg, Russia, June 11–15, 2001, pp. 1406–1410.

[76] Furuskar, A., et al., "Performance of WCDMA High-Speed Packet Data," *Proceedings of 55th VTC*, Birmingham, Alabama, May 6–9, 2002, pp. 1116–1120.

[77] Kordybach, K., and S. Nousiainen, "Radio Resource Management in WCDMA-Based Networks in Emergency Situations," *Proceedings of 18th International Teletraffic Congress*, Berlin, Germany, August 31–September 5, 2003.

[78]   Li, F. Y., and N. Stol, "A Priority-Oriented Call Admission Control Paradigm with QoS Renegotiation for Multimedia Services in UMTS," *Proceedings of 53rd VTC,* Rhodes, Greece, May 6–9, 2001, pp. 2021–2025.

[79]   Dahlberg, T., K. R. Subramanian, and B. Cao, "Soft Capacity Modeling for Third Generation Radio Resource Management," *Proceedings of International Mobility and Wireless Access Workshop (MobiWac'02),* Fort Worth, Texas, October 12, 2002, pp. 33–39.

[80]   Brüggen, T., et al., "Capacity Improvement in UMTS by Dedicated Radio Resource Management," *Proceedings of 56th VTC,* Vancouver, British Columbia, Canada, September 24–29, 2002, pp. 1284–1288.

[81]   Molkdar, D., and J. Wallington, "Capacity Benefits of node B Power Sharing in a Homogenous Circuit-Switched UMTS Network," *Proceedings of 56th VTC,* Vancouver, British Columbia, Canada, September 24–29, 2002, pp. 1590–1595.

[82]   Giuliano, R., F. Mazzenga, and F. Vatalaro, "Adaptive Cell Sectorization for UMTS Third Generation CDMA Systems," *Proceedings of 53rd VTC,* Rhodes, Greece, May 6–9, 2001, pp. 219–223.

[83]   Fraile, R., and N. Cardona, "Resource Allocation Strategies for the TDD Component of UMTS," *Proceedings of the 13th IEEE International Symposium on Personal, Indoor and Mobile Radio Communications,* Lisbon, Portugal, September 15–18, 2002, pp. 21–25.

[84]   Forkel, I., et al., "Dynamic Channel Allocation in UMTS Terrestrial Radio Access TDD Systems," *Proceedings of 53rd VTC,* Rhodes, Greece, May 6–9, 2001, pp. 1032–1036.

[85]   Kurjenniemi, J., O. Lehtinen, and T. Ristaniemi, "Signaled Step Size for Downlink Power Control of Dedicated Channels in UTRA TDD," *Proceedings of IEEE 4th International Conference on Mobile and Wireless Communications Network (MWCN 2002),* Stockholm, Sweden, September 9–11, 2002, pp. 675–679.

[86]   Zhang, G., "Call Admission Control in the Uplink for WCDMA TDD Systems," *The 5th International Symposium on Wireless Personal Multimedia Communications,* October 27–30, 2002, pp. 616–620.

[87]   Kurjenniemi, J., S. Hamalainen, and T. Ristanlemi, "UTRA TDD Handover Performance," *Proceedings of IEEE Global Telecommunications Conference,* San Antonio, TX, November 25–29, 2001, pp. 533–537.

[88]   Benson, M., and H. J. Thomas, "Investigation of the UMTS to GSM Handover Procedure," *Proceedings of 55th VTC,* Birmingham, Alabama, May 6–9, 2002, pp. 1829–1833.

[89]   Tolli, A., P. Hakalin, and H. Holma, "Performance Evaluation of Common Radio Resource Management (CRRM)," *Proceedings of IEEE International Conference on Communications,* New York, April 28–May 2, 2002, pp. 3429–3433.

CHAPTER 5

Resource Management in Systems Beyond 3G

In the previous chapters radio resource management was presented for 2G, 2.5G, and 3G networks. The common characteristic of these systems is that they refer to a specific radio access network. Resource management in systems beyond the 3G is more complicated, since it is still not clear how such systems will be realized, and moreover, because it is expected that the operators can make use of multiple access networks. This chapter presents the current status and trends toward this generation, research activities that are under way, and finally radio resource management techniques that are envisaged for systems beyond 3G.

5.1 Definition of Systems Beyond 3G: Vision and Requirements

2G cellular networks, such as GSM, are mainly used for voice transmission and they are essentially circuit-switched. Extensions of 2G networks, such as GPRS (2.5G), make use of circuit switching for voice and packet switching for data transmission. 3G networks were proposed to eliminate many of the constraints faced by 2G and 2.5G networks, such as low transmission rates and incompatibility of technologies (TDMA/CDMA) in different countries. 3G networks stand on a packet-switched technology that utilizes bandwidth much more efficiently. Today the mobile world community moves toward 4G networks, using packet-switched and IP networks.

If someone would like to define 4G, they need to perform in-depth research, as 4G takes on a number of equally strong definitions. In the simplest terms, 4G is the next generation of general purpose wireless networks that are expected to replace 3G networks sometime in the future (after 2010). At the moment, 4G simply corresponds to initiatives carried out by research and development labs to surpass limitations and deal with problems experienced with 3G.

This section presents a clear view of the future 4G world as accurately as possible. Before that there is a need to give a historical overview of the technological revolution from the 2G to the 4G systems. Furthermore, a general architecture envisaged for a 4G network is presented. Finally a detailed report regarding research from key players (forums, manufacturers, and research labs) for 4G is presented not only from an architectural point of view but also regarding its potential impact, both social and economic.

5.1.1 The Way Towards 4G Systems

The evolution from first generation analog systems (1985) to 2G digital systems (1992) was only the beginning of the digital revolution. The first generation of cellular phones was based on frequency-modulated (FM) analog technology. Most countries developed their own systems, but while these phones allowed for roaming within one region, they could not be used across different countries.

This was a very important problem in the European Union, where each country had its own standard. To address this problem, the ETSI [1] specified the first 2G digital technology, GSM. GSM was mandated in the early 1990s as the digital mobile communications technology for all of Europe. The standardization of GSM starting from Europe and spreading gradually globally was a huge success. GSM, as presented in Chapter 2, was designed around digital speech services and for low bit rate data that could fit into a speech channel. However, the main drawback of GSM was its inefficiency for data communications.

Despite the fact that GSM was dominant, worldwide roaming still presents some problems with the U.S. standard IS-95 (a CDMA rather than TDMA digital system) and IS-136 (a TDMA variant). Interim standard 95 (IS-95) [2] is a North American digital cellular standard based on the CDMA technology. It was formally adopted by the Telecommunication Industry Association (TIA) in 1993. IS-95 allows each user within a cell to use the same radio channel, and users in adjacent cells also use the same radio channel. This frequency reuse increases the efficiency of spectrum usage. IS-95a, which is often referred to as CDMAOne is capable of supporting voice and data at 14.4 Kbps whereas IS-95b is capable of supporting data rates up to 115 Kbps.

Interim Standard 136 is often referred to as digital advanced mobile phone service (DAMPS). This system uses a TDMA process over the radio interface, which is similar to that used in GSM. The IS-136 TDMA standard offers the opportunity for integrating macrocellular and indoor wireless access in a seamless fashion while using a low-cost handset that is based entirely on 30-kHz frequency division duplex channels [3].

The large customer base of these technologies will be the target group for the new services that will be launched in the 4G framework. Digital data services are well-known to the business community, but the major concern for public users is how to increase the speed on the one hand and lower the cost at which data can be handled by mobiles on the other. To address the rapid increase in the popularity of such services, GPRS has been added to GSM in 2000 to allow high-speed packet-based communications, as described in more detail in Chapter 3. GPRS is a 2.5G technology that offers data transfer rates of 56–114 Kbps. Global deployment begun in late 2000 and is ongoing. EDGE (an extension of GPRS) can support data transfer rates up to 384 Kbps. Besides that, WAP as well as i-mode are the migrations to allow higher data rates as well as speech, prior to the introduction of 3G.

The introduction of 3G has been launched by the 3G standardization activities that started in 1999 by ETSI. At the same time IMT-2000 is the ITU globally coordinated definition of 3G covering key issues such as frequency spectrum use and technical standards [4]. The most fundamental requirement for the standardization of IMT-2000 is to create a standard that provides capabilities to facilitate worldwide roaming and mobile multimedia applications as defined in Recommendation ITU-R

M.1455 [5]. Such systems extend services to high-quality multimedia (multirate) and to convergent networks of fixed, cellular, and satellite components. The radio air interface standards are based upon WCDMA (UTRA FDD, UTRA TDD in UMTS [6] and multicarrier CDMA 2000, single carrier UWC-136 on derived U.S. standards), as described in Chapter 4. The WCDMA-based air interface has been designed to provide improved high-capacity coverage for medium bit rates (384 Kbps) with limited coverage and with up to 2 Mbps in indoor environments with no mobility.

Whereas 2G operates in 900- and 1,800/1,900-MHz frequency bands, 3G is intended to operate in wider bandwidth allocations at 2 GHz. These new systems will comprise microcells as well as macrocells to deliver the higher capacity services efficiently. Thus 3G will provide a significant step in the evolution of mobile personal communications. It is true that 3G can support the multimedia type of services at improved speeds and quality compared to 2G. However, even 3G systems have limitations in the offered data rates as well as in the mobility of terminals, which needs to be seamless and uninterrupted.

To support IP-based services and QoS, new standards for broadband indoor wireless communications (micro mobility), HIPERLAN 2 (ETSI/BRAN) and IEEE 802.11a [7] have been proposed and standardized. Such systems are based on OFDM [8] rather than CDMA and are planned to operate in the 5-GHz band.

HIPERLAN/2 is a flexible radio LAN standard designed to provide high-speed access (up to 54 Mbps at PHY layer) to a variety of networks including 3G mobile core networks, ATM networks and IP-based networks, and also for private use in wireless LAN systems (ETSI definition [9]).

To summarize the major technologies Figure 5.1 depicts the status of the various systems and technologies in terms of data rates and mobility.

Please consider the following in consideration of Figure 5.1:

- Higher data rates are difficult with CDMA—due to excessive interference between services.
- Constraints are imposed on the core network by the air interface standard.
- It is difficult to provide a full range of multirate services, all with different QoS and performance requirements.
- The available bandwidth in the 2-GHz bands allocated for 3G will soon become saturated.
- Constraints on the combination of frequency and time division duplex modes are imposed by regulators.
- Services in different environments and for different user mobility are inefficient.

In response to this, 4G must be by nature dynamic and adaptable in all aspects with built-in intelligence. 4G also needs to reflect the convergence issues already mentioned, and in particular integration of the radio access and the core network elements, which must be designed as a whole rather than segmented as was the case in the past. All characteristics regarding the transition from 1G to 4G are depicted in Table 5.1.

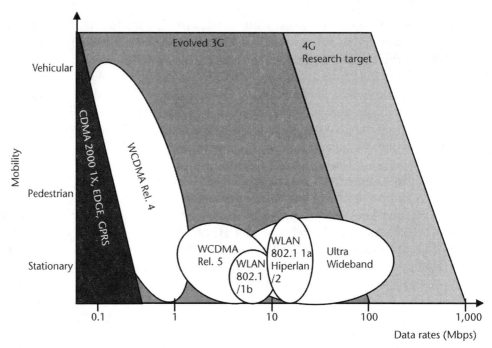

Figure 5.1   Communications systems in terms of data rates and mobility.

Table 5.1   Previous and Future Systems Characteristics

Attribute	1G	2G	2.5G	3G	4G
Starting time	1985	1992	1995	2002	2010–2012
Driven technique	Analog signal processing	Digital signal processing	Packet switching	Intelligent signal processing	Intelligent software Auto configuration
Representative standards and technologies	AMPS, TACS, NMT	GSM, TDMA	GPRS, I-mode, HSCSD, EDGE	IMT-2000 (UMTS, WCDMA, CDMA2000)	OFDM, UWB
Radio frequency (HZ)	400–800M	800–900M, 1,800–1,900M	2G	3G–5G	
Bandwidth (bps)	2.4K–30K	9.6–14.4K	171–384 K	2M	10–20M
Multiple access technique	FDMA	TDMA, CDMA		CDMA	FDMA, TDMA, CDMA
Cellular coverage	Large area	Medium area		Small area	Mini area
Core networks	Telecommunications networks	Telecommunications networks		Telecommunications networks, some IP networks	All-IP networks
Service type	Voice Mono-service Person-to- person	Voice, SMS Mono-media Person-to-person	Data service	Voice, data Some multimedia person-to-machine	Multimedia machine-to-machine

5.1.2   General Architecture

Several 4G concepts that are currently under discussion could find their way in a common approach by convergence of those definitions. Generally speaking everybody agrees that 4G network is the name given to an IP-based mobile system that provides access through a collection of radio interfaces [10]. A 4G network promises seamless roaming/handover and the best connected service, combining multiple radio access interfaces (such as HIPERLAN, WLAN, UMTS, Bluetooth, and GPRS) into a single network that subscribers may use. With this feature, users will have access to different services, increased coverage, the convenience of a single device, one bill with reduced total access cost, and more reliable wireless access even with the failure or loss of one or more networks.

From another point of view, a 4G architecture will include five basic areas of connectivity: personal area networking (such as Bluetooth), local high-speed access points on the network including wireless LAN technologies (such as IEEE 802.11x and HIPERLAN), national coverage through cellular coverage of 2–3G networks (such as GSM, GPRS, UMTS, and EDGE), regional connectivity under technologies such as DVB-T, DAB, and satellite global coverage. Under this umbrella, 4G calls for a wide range of mobile devices that support global roaming and high data rates. Each device will be able to interact with Internet-based information that will be modified on the fly to comply with the characteristics of the network being used by the device at that moment. These characteristics are imposed by the idea of pervasive computing [11]. In Figure 5.2 the 4G vision is presented in terms of addressed technologies and their mobility support.

The glue for all this is likely to be the software-defined radio (SDR) technology. SDR enables devices such as cell phones, PDAs, PCs, and a whole range of other devices to scan the airwaves for the best possible method of connectivity, at the best price. In an SDR environment, functions that were formerly carried out solely in

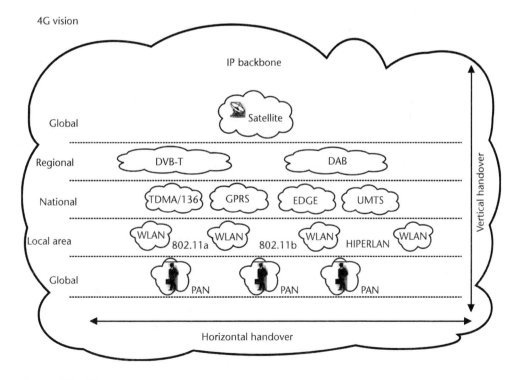

Figure 5.2   4G vision.

hardware—such as the generation of the transmitted radio signal and the tuning of the received radio signal—are now performed by software. Thus, the radio devices are made programmable and able to transmit and receive over a wide range of frequencies while emulating virtually any desired transmission format.

5.1.3 Visions About 4G

If someone goes through a quick search on the Internet about what 4G is and how could be developed in next years, he or she comes up with a quite large number of different answers. In a widely adopted definition, as of the first half of 2002, 4G is a conceptual framework for or a discussion point to address future needs of a universal high-speed wireless network that will interface with the IP backbone network seamlessly. In other words 4G represents the hope and ideas of a group of researchers in various standardization forums, among worldwide mobile manufacturers and other infrastructure vendors that must respond to the increased demands from the mobile users. Section 5.1.4 provides the view of the most significant forums related to 4G and furthermore the approach of some of the biggest manufacturers in the mobile world. This approach will provide a more comprehensive view of 4G is going to be.

5.1.4 Standardization Bodies and Forums

Standardization is one of the key issues in the research activities for next generation wireless systems. The development of a proactive strategy for 4G in standardization bodies and forums provides valuable starting conditions for the marketplace to best benefit from the corresponding findings and results for next generation systems. Those wishing to have a global view of future 4G systems should emphasize the standardization activities and discussions that take place in the forums. Bodies and forums such as the IMT-2000, 3GGP, 3GPP2, 4G Forum, WWRF, and IPv6 Forum give a satisfactory overview of 4G systems.

IMT-2000 is the global standard for 3G wireless communications, defined by a set of interdependent ITU recommendations. IMT-2000 provides a framework for worldwide wireless access by linking the diverse systems of terrestrial and satellite-based networks. It focuses on exploiting potential synergies between digital mobile telecommunications technologies and systems for fixed and mobile wireless access [12]. The working group WP 8F is responsible for the overall radio system aspects of IMT-2000 and beyond. It is responsible within ITU-R Study Group 8 for issues related to the terrestrial component of IMT-2000 and beyond [13]. The activities of WP 8F consist also of the long-term vision of spectrum needs, investigation of evolving marketplace needs for 4G communications, higher data rate capabilities, and overall for the worldwide advancement of IMT-2000 systems into 4G systems.

The 4G Mobile Forum (4GMF) is the first international technical body dealing with the next generation broadband wireless mobile communications, which encompass wireless access, wireless mobile, WLAN, and packet-division-multiplexed (PDM) networks [14]. The technological view of 4GMF is that an integrated 4G mobile system will provide wireless users with an affordable broadband mobile access solution for secured and scalable wireless mobile services and applications with value-added QoS through all the way from the MAC layer up to application layer. From the economic prospective point of view, 4G Forum predicts that the business of 4G will be extremely high. In particular, by 2008, this 4G mobile market will be over $400 billion [14].

The Wireless World Research Forum (WWRF) is a global organization in which members are manufacturers, network and service providers, R&D centers, small and medium enterprises (SMEs), and universities. Beyond many other objectives, WWRF's main goal is to generate, identify, and promote the research areas for mobile and wireless systems toward a future wireless world [15]. WWRF's technical view for future systems is put into a much wider way rather than focusing in a purely "technical" discussion based on network concepts and radio interfaces. WWRF looks at a user-centric approach, definition, and development of new services and applications that become possible with next generation systems and the need for new business models that may prevail in the future [16]. Inside WWRF it is believed that the development of next generation communication systems will be truly user-centric. Therefore 4G has to address communications among humans, things, and cymans (synthetic counterparts in a virtual cyber world) [16]. Achieving these new opportunities will be possible, especially in the close area of the human mobile user, as WWRF states.

Nobody argues that IP will play a major role in the future networks. Almost all visions come to a common understanding that 4G will be an IP-based mobile communications system. The transition from existing to next generation systems will inevitably require an IPv6 update for all the IP networks. IPv6 is a new version of IP that is designed to be an evolutionary step from IPv4. It is a natural increment to IPv4. IPv6 is designed to run well on high-performance fixed networks and at the same time still be efficient for wireless systems [17]. For sure there will be a transition period between IPv4 and IPv6 networks, but IPv6 will provide the long-term solution. IPv6 Forum's [18] main objective is to promote IPv6 by dramatically improving the market and user awareness of IPv6, creating a quality and secure next generation communication system, and allowing worldwide equitable access to knowledge and technology, embracing a moral responsibility to the world [18]. Enhanced IPv6 networking will be needed to allow seamless mobility within and between administrative domains and systems, to provide an end-to-end QoS architecture, and for location-based communication and location information management. Mobility is the most compelling reason for rapidly moving toward IPv6 [19].

The IPv6 working group within IPv6 Forum was originally chartered as the IP next generation (IPng) working group to implement the recommendations of the IPng area directions as outlined at the July 1994 IETF meeting and in "The Recommendation for the IP Next Generation Protocol," RFC1752, January 1995. The Internet Engineering Task Force's (IETF) IPv6 working group recently published (September 2002) the document that contains the recommendations for IPv6 to the 3GPP [20].

3GPP is a collaborative agreement that was established in December 1998. The collaborative agreement brings together a number of telecommunications standards bodies, which are known as "organizational partners." The current organizational partners are ARIB, CCSA, ETSI, T1, TTA, and TTC [21]. 3GPP finalized successfully WCDMA based on IMT-2000 standard and set it as Release 99. UMTS/3GPP

Release 99 has already been accepted by major operators in Europe and Japan. Commercial services have been developed both in Japan (2001) and Europe (2002). Now 3GGP moves fast toward IP-based network evolution based on a systematic approach for service provisioning, enhancing the relation with ITU. Moreover it enforces the framework for 4G evolution based on an all-IP paradigm.

The 3GPP2 initiative is the other major 3G organization. 3GPP2 is a collaborative 3G telecommunications specifications-setting project comprising North American and Asian interests developing global specifications for ANSI/TIA/EIA-41 cellular radiotelecommunication intersystem operations network evolution to 3G [22]. It promotes the CDMA2000 system, which is also based on a form of WCDMA technology. The major difference between 3GPP and 3GPP2 approaches is that of the air interface specified. 3GPP has specified a completely new air interface without any constraints from the past, whereas 3GPP2 has specified a system that is compatible with CDMA (IS-95 standard), the 2G system implemented in the United States and elsewhere in the world. In North America, IS-95 systems are already using frequency bands allocated for 3G systems. CDMA2000 1x and 3x were released with guaranteed backward compatibility. Commercial services in the United States are expected to start in 2005. Migration of all legacy systems and services to 4G (all-IP networks) is increasing the need for common specification parts between 3GPP and 3GPP2.

5.1.5 Network and Device Manufacturers

In the long run, 4G wireless systems will be developed and begin their operation. Advanced features of wireless mobile systems, including high data rates for multimedia applications, global roaming capability, and coordination with other network structures, will enable the introduction of these applications and services that could not be launched with the existing wireless mobile systems [23]. 4G represents also the future view of a large number of researchers of key network and device manufacturers as well as infrastructure vendors that must prepare to provide the required infrastructure for the envisaged applications. The prospects and plans for the mobile future from the key manufacturers in the mobile world community are presented in this section.

Manufacturers should provide a wide variety of leading-edge mobile 4G devices. Already many of them are moving toward the design of the future 4G systems. As an example, NTT DoCoMo in Japan has been conducting research on 4G mobile communications since April 1998 [24]. In the context of this research, DoCoMo has conducted indoor trials on an experimental system that incorporates base station and mobile station equipment to evaluate key technologies in 4G packet wireless access and to demonstrate its benefits by employing the implemented experimental system [24]. Recently, this organization announced that it has succeeded a 100-Mbps downlink and 20-Mbps uplink transmission experiment in an indoor environment using an experimental system for 4G mobile communications [25].

DoCoMo 4G view could be summarized, in five key letters, "M.A.G.I.C," the scenario that will be the background for the 2010 mobile technology. This scenario consists of *m*obile multimedia, available *a*nytime, anywhere and to anyone, with *g*lobal mobility support and offering *i*ntegrated wireless solutions that meet user's needs and support user's daily activities with customized information according to each person [26].

Furthermore, engineers around the world, like at Sun Microsystems Laboratories [27], are working on building wireless technologies that promise to integrate voice and Web data in an IP-based mobile communications system [28]. According to Sun Microsystems, 4G is all about an integrated, global network that is based on an open systems approach. The goal of 4G, according to them, is "to replace the current proliferation of core cellular networks with a single worldwide cellular core network standard based on IP for control, video, packet data, and VoIP."

Sun Microsystems presents a 4G wireless network as an all IP-based network that will have intrinsic advantages over its predecessors. Sun supports that IP gives full flexibility, independent from the access technology (e.g., 802.11, WCDMA, Bluetooth, HYPERLAN/2, CDMA, etc.). "The core [IP] network can evolve independently from the access network. That's the key for using all IP," Sun's engineers state [28]. From an economical aspect, Sun Microsystems defends that a 4G IP wireless network will enjoy a financial advantage over 3G as well because equipment costs are believed to be four to 10 times cheaper than the equivalent circuit-switched equipment for 2G and 3G wireless infrastructures.

Moreover, Motorola engineers underline the fact that while 3.5G will be an evolutionary system, the vision for 4G will require new air interface technology, new spectrum allocations, and extensive research, including new modulation schemes [29]. In addition, Motorola supports that 4G is expected to have beneficial results on technological aspects such as dramatic improvements in data transmission (as much as 10 to 50X over 3G), dramatically lower costs, and services that 3G will not be able to provide, such as high-quality full-motion video, high-speed mobile Internet access, and mobile video on demand [29].

At Ericsson premises [30] there is a belief that 4G wireless technology research is already needed because it will simply succeed the 3G one. The aim for 4G, Ericsson supports, is 100 Mbps and the time of being applied to the public services is at the Olympic Games of 2012. Based on their statement it would put together personal devices that can connect to each other without cables, such as wireless headphones, a personal data organizer (PDA), and a mobile phone, with devices working in a wireless local area computer network, such as laptops and wide area networks, such as 3G telephony. The trick, however, for Ericsson will be to make them all work together without glitches [31].

Likewise, Nokia's view [32] about next generation systems is that 4G is a research topic for air interfaces and systems to be considered after 2010. The major driver for starting the research for 4G, according to Nokia, is radio performance and higher throughputs; data rates that could reach up to 5 to 10 bits/s/Hz for local efficiency and much more than 1 bits/s/Hz for wide area multicell efficiency are envisaged. Nokia holds that the CDMA system may provide a better spectrum efficiency for wide area coverage, while the TDMA system may provide this effectively for local area coverage. Therefore, Nokia believes that 4G's major objective will be high data rates everywhere in combination with hyperavailability of all media [32]. Some of the key technology issues to achieve these requirements are the multicarrier appliance, which allows flexibility in the use of the available spectrum; spatial multiplexing (MIMO) [33], which achieves high spectrum efficiency; and the investigation of multicarrier CDMA and nonspread multicarrier TDMA, which help to determine the best operating range [32].

5.2 Initiatives and Research Activities Toward Systems Beyond 3G

This section provides an overview of the initiatives and research activities that are already under way toward enabling seamless and efficient communication in heterogeneous wireless networking environments. The approaches described herein are based each one on different assumptions, and they aim at fulfilling different needs; however they are important first steps for the definition and deployment of the 4G in mobile communications. Emphasis is put on the corresponding resource management strategies and schemes that are proposed or adopted. We have to mention also that there are currently more than 1,000 projects worldwide that are tackling systems beyond 3G issues. On the other hand even precommercial work is taking place. For example NTT DoCoMo in Japan has announced from March 2002 trials of the so-called 4G I-MODE technology, which aims at providing high-speed wireless communication on the move with up to 20-Mbps data rates. Of course in the review presented herein we will present just a representative set of initiatives, which we believe will provide satisfactory coverage.

5.2.1 The IST Project CAUTION++

A hybrid 3GB system is examined in CAUTION++ framework aiming at the efficient interworking of four different access networks (GSM, GPRS, UMTS, and WLAN) under a unified and hierarchical resource management model. CAUTION++ is an Information Society Technologies (IST) project of the European Commission that started in November 2002 and has 30 months duration. The requirements for the system's definition are examined from three viewpoints: end user, network operator, and service provider [34].

Based on the extracted requirements, a hierarchical radio resource management system is defined in CAUTION++, which is composed of three main components, linked by means of dedicated wire lines or an IP-based backbone network. These components are described as follows:

- The interface traffic monitoring unit (ITMU) extracts the status information from each one of the access networks by providing the traffic information and the alarm message type toward the corresponding RMU entity.
- The resource management unit (RMU) is the core element for each network separately, where the management techniques are decided and executed.
- The global management unit (GMU) is the centralized CAUTION++ core component that enables the coexistence of different access networks, providing a global resource management to them and further assistance for the correct operation of the RMU tasks.

The RMU performs local resource management of each single network resource whereas the GMU is responsible for global resource management as well as for interworking coordination and optimization. Moreover, due to the fact that different access networks are located in the same area with either cells or access points, a location server (LS) is also used to provide location information about the mobile users to the GMU for maximizing the efficiency of its operations.

The CAUTION++ system provides the framework for unified resource management, based on the "monitor and manage" concept. Resources at the air interface are real-time–monitored, and a centralized hierarchical system receives alarms from the distributed monitoring components, so that a set of management techniques is selected and applied where and when needed to keep up a satisfactory level of the available resources, the QoS, and the interoperation between different wireless systems. Forced vertical handover of the user (i.e., from WLAN to GPRS) can be performed in normal congestion. However, when the operator of the superset of access networks is unable to decongest a phenomenon in its network, the CAUTION++ system provides even the capability to perform a forced vertical-vertical handover to another operator's supernetwork (with available resources) to finally avoid the user's drop [35].

5.2.2   The IST Project MONASIDRE

The MONASIDRE project bases its work on the assumption that in the future, UMTS, HIPERLAN-2, and DVB-T can be three cooperating components of a heterogeneous radio environment that offers wireless access to broadband IP-based services [36]. The main objectives of this project are: (1) to monitor and analyze the statistical performance and the associated QoS levels provided by the network elements; (2) to support interworking with service provider mechanisms, so as to allow service providers to dynamically request the reservation or release of network resources; (3) to perform dynamic network planning as a result of resource management strategies to optimize delivery of services to mobile users under a spectrum-limited constraint; and (4) to adequately map the IP-based network resources to the radio resources.

The resource management process is divided in two main steps: (1) splitting of traffic over the available networks, and (2) management of the split traffic (splitting proposal) into the specific networks. The first step takes into account the load of each network in the heterogeneous network environment, the users' profiles, the cost of the specific services, and the desirable QoS and finally returns a traffic-splitting proposal. The result of the second step, which controls the split traffic (proposal of the first step) into each network (UMTS, HIPERLAN-2 AND DVB-T), is a set of parameters that configure the network elements of each environment.

5.2.3   The IST Project COMCAR

The main focus in COMCAR is on asymmetrical and interactive IP-based services. The radio technologies and infrastructures used in COMCAR are GSM, UMTS, digital video broadcast (DVB), and digital audio broadcast (DAB), and the purpose is to bring asymmetric high-quality IP-based services to vehicles like cars and trains [37]. COMCAR provides a flexible communication environment in which QoS parameters change on a wide scale. The project examines how this scenario might influence emerging new Internet technologies for integrating QoS in wireless IP networks. Furthermore, COMCAR develops mobile middleware technologies that allow adaptive mobile multimedia applications to react to a changing user context.

The COMCAR solution offers a way to coordinate the spectrum's usage for mobile communication systems through a mechanism called System spectrum coordinator (SSC). A very useful component for coordination between different systems as part of the system spectrum coordinator is the common coordination channel (CCC). The purpose of this logical channel is to transfer information on provided services, supported spectrum, and traffic characteristics like load and data rate capabilities to the terminals. The CCC in COMCAR supports a flexible, location-dependent allocation of spectrum to different systems (e.g., spectrum allocated in Sweden for DVB-T could be used in Germany for UMTS). A terminal could read from a CCC of a member system, where to find the different systems in the spectrum. In this respect the CCC makes it easier to use a particular system globally since a unique spectrum allocation in the world for the system is not required anymore [38].

5.2.4 The IST Project Dynamic Radio for IP Services in Vehicular Environments (DRIVE)

The overall objective of DRIVE project is to enable spectrum-efficient high-quality wireless IP in a heterogeneous multiradio environment for delivering in-vehicle multimedia services enabling universal access to information and support for education and entertainment [39]. Furthermore DRIVE designs location-dependent services that adapt to the varying conditions of the underlying multiradio environment. To achieve these objectives DRIVE addresses two key issues: (1) interworking of different radio systems (GSM, GPRS, UMTS, DAB, and DVB-T) in a common frequency range with dynamic spectrum allocation and (2) cooperation between network elements and applications in an adaptive manner.

DRIVE's architecture treats the different radio systems independently and converges them at the network layer. The transport of high-end mobile multimedia services has to cope with varying radio resources and network capabilities. The applications must be able to adapt their behavior to the current radio environment and the available resources. Scalable and modular applications that can be distributed and dynamically downloaded to different physical entities can actively support the requested services. Here network elements and applications cooperate in an adaptive manner.

A traffic control (TC) function is responsible for coordinating the traffic distribution process and selecting the appropriate RAN for a specific service. The traffic control function has as inputs the terminal-related information (such as terminal capabilities and location), the user preferences (preferred access system and cost preferences), traffic parameters, QoS requirements, and the status of the network (e.g., load and current capacity). The RAN selection takes place at the start of a session and when a change in the network status or radio conditions occurs.

DRIVE also tackles the issue of dynamic spectrum coordination and allocation. This requires methods for automatic spectrum selection, techniques for adaptation of sender and receiver to different frequencies, and investigations to assure the coexistence (with regard to interference and sensitivity) of different radio technologies. In a scenario of time- and location-dependent availability of spectrum one alternative to organize the dynamic spectrum allocation is to define a (logical) CCC as the one referred to in the COMCAR project. The vision of DRIVE is that any operator could work with any radio system on any band.

5.2.5   The IST Project OVERDRIVE

The OVERDRIVE project aims at UMTS enhancements and coordination of existing radio networks into a hybrid network to ensure spectrum efficient provision of mobile multimedia services [40]. IPv6-based architecture enables interworking of cellular and broadcast networks in a common frequency range with dynamic spectrum allocation (DSA). The project's objective is to enable and demonstrate the delivery of spectrum efficient multicast and unicast services to vehicles. In particular OVERDRIVE is focusing on the following:

* Improving spectrum efficiency by system coexistence in one frequency band and DSA;

* Enabling mobile multicasts with UMTS enhancements and multiradio multicast group management;

* Developing a vehicular router that supports roaming into the intravehicular area network.

DRIVE concentrates on the evaluation and investigation of the contiguous DSA scheme, where contiguous blocks of spectrum are assigned to different RANs, separated by suitable guard bands. If a RAN (e.g., UMTS) requires additional spectrum, this should be taken from the RAN in the adjacent spectrum (e.g., DVB-T) and vice versa, whereby the guard band is shifted accordingly, maintaining the same width. This results in limited flexibility for spectrum partitioning. OVERDRIVE overcomes these limitations by allowing more flexible DSA schemes. Moreover, the employed hierarchical architectures require investigations for the regional coexistence of different cell-layer access systems in a single shared frequency band. OVERDRIVE pursues analytical and simulation based studies for advanced DSA methods, such as fragmented and cell-by-cell DSA.

5.2.6   The IST Project Broadband Radio Access for IP-Based Networks (BRAIN)

The project BRAIN provides a broadband multimedia IP-based radio technology by supporting several high data rate users via one base station (e.g., provision of 2 Mbps for 10 users such that the total data rate is around 20 Mbps per RF channel). It also supports a wide range of services (point-to-point, point-to-multipoint, symmetric, and asymmetric) and allows roaming as well as inter- and -intrasystem handover between GSM/GPRS and UMTS networks. The following list presents the objectives of BRAIN [41]:

* To facilitate the development of seamless access to existing and emerging IP-based broadband applications and services for mobile users in global markets;

- To propose an open architecture for wireless broadband Internet access to allow an evolution from fixed Internet, emerging wireless/mobile Internet specifications, and UMTS/GSM;
- To facilitate new business opportunities for operators, service providers, and content providers to offer high-speed (up to 20 Mbps) services complementary to existing mobile services;
- To contribute actively to global standardization bodies in the necessary time scales to impact significantly the international standardization.

The BRAIN access network is based on end-to-end IP for all real-time and nonreal-time services in public and private networks. The IP-based core network connects all involved radio access schemes of the BRAIN architecture. BRAIN is applied in picocells as well as in-building and home cells and in urban and suburban cells. Full coverage with reduced data rates is being provided by GSM/GPRS/EDGE and UMTS. These systems complement each other depending on the different radio environments and service needs.

Regarding resource management, in the BRAIN architecture the design goal was to maintain a clean separation between network-related issues (which should be generic to any air interface) and issues specific to a given air interface. In the BRAIN system, the network layer has to take part in any procedure that involves handovers between access routers (also called network layer handovers), since this requires rerouting and possibly network-related QoS reconfiguration within the wired access network. This makes radio resource management at least partially the concern of the network layer. Note that the BRAIN RRM function requires an interface to the traffic engineering/management functions within the wired access network, and to QoS brokers within the IP core, to complete the admission control decision.

5.2.7 The IST Project Mobile IP-Based Network Developments (MIND)

MIND, IST-2000-28584, is a follow-up project that develops and validates the concepts reported in BRAIN. MIND provides a further evolution to allow ad hoc networks and PANs leveraging the public infrastructure to gain access to the Internet, to long-distance calls facilities, and to services, but foremost to security support. For providing user perceived QoS management, it provides a QoS framework that enables distributed multimedia applications with an adaptation mechanism and QoS provision on an end-to-end basis, across heterogeneous wireless access networks. Within MIND, the main objective of the networking work was to enhance the access routers and the mobile hosts to operate in a self-organizing wireless environment supporting ad hoc mode.

Regarding resource management in the MIND ad hoc network environment, there are a number of additional issues that have to be considered. First, because of the dynamic topology and instability of the network, it is difficult to support a centralized RRM model (where one node in the network controls the allocation of all radio resources), and therefore the RRM algorithm is distributed across multiple nodes. For that some form of signaling is required to distribute the information required by the RRM algorithm (e.g., local resource availability between the different instances of the distributed RRM algorithm running on multiple network nodes).

Since the environment consists of heterogeneous link layer technologies, some aspects of the RRM algorithm are required at L3 to support the exchange of information between nodes in a technology neutral (and possibly multihop) manner. Furthermore, in a multihomed environment, resource usage across different interfaces based on different link layer technologies is required. The MIND RRM investigation determined to split the functionality between L3 and L2 to support the exchange of parameters between the RRM entities residing at the different layers. The resultant architecture defines the location of the different aspects of RRM, and interlayer communication. For interlayer communication, it is proposed that a single function is used as a "gateway" between the layers.

This implies that the RRM architecture has controlling entities at both L2 and L3 that coordinate the RRM functionality between the layers allowing close cooperation without tight integration. This approach achieves good overall system performance. The RRM architecture in MIND contains a measurement entity for performing monitoring of the air interface, a resource manager (RM) entity for allocating resources, an admission control (AC) entity for controlling access to those resources, queue and scheduler entities, and an output conditioner entity, which conditions the data flow to meet the current radio link constraints.

5.2.8 The IST Project Configurable Radio with Advanced Software Technology (CAST)

CAST is a characteristic example of a project that deals with radio resource management using the SDR concept. The main scope of the project was to research and demonstrate an intelligent software-defined architecture that is reconfigurable to operate in both GSM and UTRA environments. Software radio is a generic term for communicating devices whose behavior can be redefined by software. The CAST project focuses on intelligent reconfiguration of mobile networks by adding extra components to the existing mobile network. Many elements of the mobile network (elements like MS, BS, and MSC) are utilized within the project and other components of the system are emulated. The whole software radio concept is implemented using JAVA virtual machines [42].

In the CAST project resource management is handled by a resource controller, which is a distributed software module. The resource controller's task is to dynamically set up and maintain processing chains to ensure the required performance. The managed processing chains consist of functions that are placed onto reconfigurable hardware objects, such as DSPs and FPGAs. The resource controller provides a map of the resources available in the radio network elements and allocates generic resources for tasks. It also schedules and downloads suitable software modules for each processing resource when required. The resource controller software is a collection of algorithms and methods that allow an efficient hardware configuration. This process involves intelligent mechanisms for locating suitable hardware capacity and choosing the appropriate configuration. During the reconfiguration process the decisions are made based on different aspects, which it is possible to vary from system to system.

5.2.9   The Next Generation Wireless Internet Project

This is a project that runs at the Georgia Institute of Technology. The scope is to investigate several key elements that are necessary to realize a protocol stack for a mobile station in 4G wireless systems [43]. The ultimate objective is to develop a protocol suite that would adapt itself to different aspects of heterogeneity exhibited by next generation wireless systems. A new transport control protocol is proposed for nonreal-time data traffic over new generation wireless networks. The objective of the protocol is to achieve high throughput performance in the heterogeneous environments of new generation wireless networks. Also for adaptive real-time applications, a new rate control scheme is introduced, and a new routing algorithm is proposed for multitier, multinetwork terrestrial routing in next generation wireless networks.

5.2.10   The 4G Mobile Network Architecture and Protocols Project

This project is carried out by WINLAB at Rutgers University. This project aims to explore the fundamentals of next generation mobile network architectures and protocols, looking beyond the issues addressed by today's mobile IP, WLAN, and 3G solutions [44]. In particular, it investigates an open-architecture, programmable mobile network approach that permits gradual evolution of service features via ad hoc peer-level collaboration of wireless network entities, potentially reducing the need for complex standards that anticipate all future needs. An experimental open-architecture 4G mobile network test bed is being set up to evaluate different approaches both in terms of protocol functionality and software performance.

In particular the following protocol design scenarios are being investigated:

- Compatible upgrades to WLAN protocols for service features such as flow QoS and multicasting;
- Interworking (e.g., global roaming and handoff) of multiple radio link technologies such as Bluetooth, 802.11, GPRS and 3G/WCDMA;
- Self-organizing ad hoc network protocols for discovery and routing, with a particular focus on a hierarchical 802.11b architecture consisting of sensor nodes (SNs), radio forwarding nodes (FNs), and access points (APs);
- Content delivery techniques for mobile users, including those based on proactive Infostations caching and novel semantic routing techniques.

5.3   Radio Resource Management in Systems Beyond 3G

In the previous chapters radio resource management for the various generations of cellular wireless systems was presented. Focus was given on monitoring the KPIs and on resource management techniques that can be applied to reconfigure the network dynamically and increase its performance. This section provides a specific approach for managing resources in systems beyond 3G.

As mentioned in the introduction, one of the ways to achieve continuous and efficient radio resource management is to monitor the KPIs of the systems and enable alarm triggering and dynamic reconfiguration when the system is congested. Reconfigurability is envisaged not only in the terminals but also in the serving networks and the services they provide. It is expected that reconfigurability will enhance the network management functionalities, by allowing for much more flexible and dynamic resource management [45, 46]. This concept can be applied for any wireless system and for heterogeneous radio environments as well.

Figure 5.3 shows the architecture of a system that monitors several network segments and supports efficient decision-making based on predefined business models. This architecture aims to manage the resources of each system separately and enable system cooperation for efficient capacity management [47].

Two management subsystems, one for real-time monitoring of each network segment and one for resource management, comprise the core of such a system. The resource management subsystem consists of two additional elements, namely the resource management and the global management element, responsible for local and global RRM respectively.

5.3.1   Monitoring Requirements

In heterogeneous radio environments, like the ones presented in the previous sections, a multielement management system should monitor traffic, predict and recognize shortcomings, and react to network congestion situations, trying to solve congestion in overloaded sectors of wireless networks. If we consider 4G as a number of coexisting wireless networks, including GPRS, UMTS (IMT-2000), WLAN or any other wireless system, the traffic monitoring system should be distributed, since the reports that will be used are generated in each affected area, including the MSC, SGSN, and RNC. It is feasible to make use of universal reports for the real-time monitoring. Of course there is a mapping of the specific reports with the typical reports, according to the ETSI/3GPP recommendations. The monitoring subsystem will perform the monitoring and forward alarm messages to the centralized resource management subsystem, when congestion is detected [47, 48]. The monitoring should be based on the KPIs that were presented in the previous chapters.

For GSM the major KPIs are the following:

- Traffic;
- Call set up success rate;
- Handover SUCCESS RATE;
- SDCCH blocking rate (SDCCH BR);
- TCH blocking rate;
- RACH/AGCH/SDCCH/TCH/PCH utilization.

For GPRS, the KPIs that characterize the network's performance are the following:

- Reliability data connection success rate/session release failure rate;

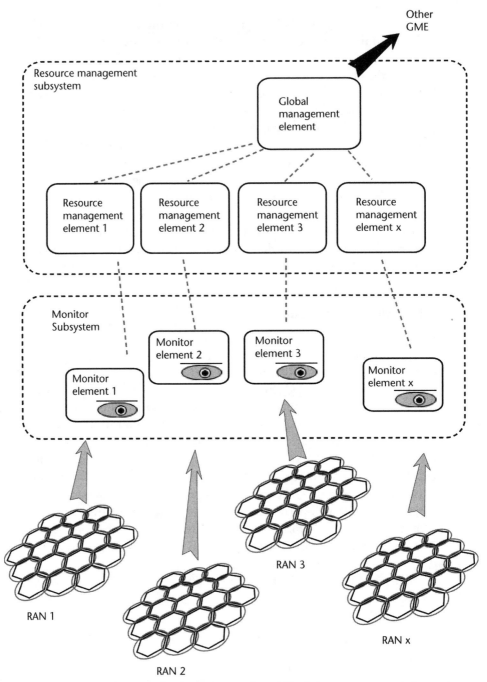

Figure 5.3 System architecture for dynamic reconfigurability in heterogeneous wireless systems.

- Delay;
- Mean/peak throughput;
- GPRS utilization;
- Attach attempts failure ratio.

WLANs will be an important part of 4G; therefore, KPIs for this kind of a network should also be defined. The major indicators are the following:

- Latency (delay);
- Jitter;
- Availability;
- Errors;
- Packet loss;
- Average/peak user data throughput.

Finally, concerning UMTS, the following KPIs should be monitored:

- Delay;
- Peak/mean throughput;
- Throughput change/total throughput;
- Service quality (SQ);
- Hard/soft blocking;
- Hard/soft dropping;
- TrCH BLER;
- Bad quality time;
- TPC tendencies;
- Allocated codes;
- DL TX power;
- UL interference;
- Neighbors' load;
- CPICH RSCP;
- Time utilization;
- SHO overhead;
- Location update.

For each of the above KPIs, two thresholds should be defined, namely the upper and the lower one. The upper one is the threshold that is defined after studying the network behavior and determines the stability limits of the network. Whenever this threshold is reached an alarm should be generated and forwarded to the resource management subsystem to tackle the problem. After congestion is over, usually after applying resource management techniques, the alarmed KPI should have a value less than the lower threshold.

5.3.2   Radio Resource Management Requirements

As it was mentioned in the previous section the resource management subsystem consists of the two elements, namely the resource- and global-management element, responsible for the local and global RRM. The local RRM is the functionality of the platform that executes the resource management techniques that can solve the traffic overload problem within the same RAN. For that reason resource management techniques, like the ones presented in Chapters 2–4 are required to solve the congestion problem. Depending on the wireless technology (GSM, GPRS, UMTS, WLAN) some of them are: cell breathing, admission control, handover adjustment, dynamic cell reconfiguration, and power control.

The global resource management differentiates from the local one in that it exploits the whole heterogeneous environment to provide efficient resource management. Whenever the local RRM entity is not able to reconfigure the system in such a way that solves the problem, it escalates to the highest hierarchy level that is responsible for the joint RRM. This involves vertical handover, as well as roaming of users from one service provide to another, in the case that two service providers have a service-level agreement.

5.3.2.1   Local Radio Resource Management

Whenever an alarm is received, the resource management element is responsible for identifying the traffic load scenario, based on an evaluation of all KPIs for that cell and the traffic load in all adjacent cells. By determining the overload in adjacent cells, it is possible to decide about the size of congestion. After the traffic load scenario is identified, the appropriate management techniques for that situation can be selected by means of a self-trained decision-making system. These techniques can be one of the RRMs presented in the previous chapters for each system separately. The resource management subsystem performs the decision-making of the system and finally executes a command to the NMS for congestion management. This system reconfigurability results in a "closed loop" in the architecture. The reconfigurable system is part of the feedback of the cellular network that performs the automatic control to result in a more stable state.

The resource management element may use a case-based reasoning (CBR) method that selects the appropriate management technique by matching the current situation with other situations saved in a database [49]. Subsequently this can be optimized by means of a set of fine-tuning mathematical algorithms. Starting from human knowledge about the possible traffic load scenarios and suitable resource management techniques, the resource management element builds over time a database of observed congestion cases and applied techniques and improves its performances by learning from past experiences. Searching the database of past cases makes it possible to find similar situations that occurred previously and to immediately identify the most successful strategy to deal with the current congestion event.

Once a technique has been selected to deal with a congestion event, the parameters necessary to instantiate the technique can be either obtained by similar application cases stored in the database, or can be computed with a fine-tuning model-based approach that optimizes the resource management technique to meet the network operator goals for the current scenario.

5.3.2.2   Global Radio Resource

The third component of the management hierarchy is an element that should have a coordination role in the general architecture. It communicates with all resource management elements of the lower level of the hierarchy within the resource management subsystem so that access and resource management is achieved considering all available network segments. The coordination element controls traffic rerouting as well as the seamless handover from one system to another. The various wireless access technologies have different mobility management techniques and protocols, each designed specifically for a certain network. The global management element retrieves location-related information, and the mobility management procedure may, for example, detect that a WLAN/cellular multimode terminal is approaching an area with WLAN coverage. In this case the traffic to and from the mobile terminal could be directed through the WLAN.

In addition it is responsible to choose when and under what conditions a system reselection must be done by the MS. It aims to coordinate different wireless networks by collecting changes and rationalizing interservice manipulations between systems. Data services of cellular networks will be alternatively provided by hotspot high-data rate WLAN systems in a dynamic and seamless manner. To avoid an unnecessary search of WLAN beacons, the mobile terminal should be aware of the whereabouts of the overlay system to be visited.

One of the most important aspects of the presented reconfigurable architecture is the capability of routing traffic of foreign networks. Therefore, apart from the management role within the same operator, it will act as a gateway to other operators and obviously to all kinds of networks that they have under their responsibility. As shown in Figure 5.4 there is a need for one such element for each operator involved in the hierarchy.

5.3.3   Other Requirements

In addition to the network monitoring and resource management described in the previous paragraphs there are other requirements that should be met to achieve a reliable and efficient RRM platform:

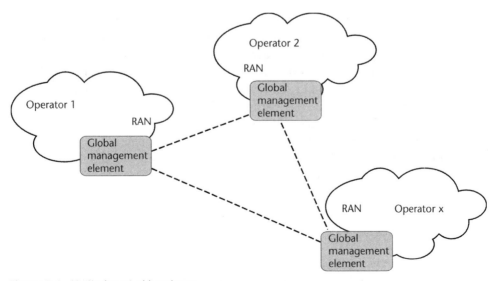

Figure 5.4   Vertical-vertical handover.

- *Information retrieval:* Obviously the system, as it is described up to now, contains components that possess useful information, such as real-time network monitoring data. This data can be exploited from different user categories, such as subscribers, technicians, and authorities. For each of these categories the KPIs can be interpreted in different ways, such as pricing, network availability, and QoS.

- *Emergency situation:* Since the proposed RRM system is able to reconfigure the system spontaneously, it should be able to prioritize users in emergency situations and guarantee service provision.

- *Manual reconfiguration:* Another feature of the RRM platform should be the manual system reconfiguration. It has been proven that in several cases the operator wishes to change some network parameters without considering the overall performance impact. Therefore, the platform should enable manual system reconfiguration.

- *Black spot:* Wireless systems of 3G and beyond will have the feature of cell breathing, which is a result of the constant power control in the downlink. This might result in black spots in the areas of coverage. For that purpose, the platform should be able to detect any black spot and offer adequate resources utilizing resources from the rest of the network segments that it is controlling.

- *Seamless service provision:* Finally, seamless service provision is required, and this can be achieved utilizing both the monitoring part of the system and the global RRM entity. Whenever traffic overload is detected, it can also be interpreted as a degradation of QoS, and the global RRM entity can force-handover the user to adjacent cells or networks to guarantee the required QoS.

5.3.4 Concluding Remarks

This chapter provided an overview of 4G future generation systems. The ideas and opinions of key technological drivers, aiming to keep up with this development are described and analyzed. We have also presented a possible architectural framework for tackling the problem of resource management in such networking environments. Definitely, 4G systems are still in formative stages. Searching around standardization bodies, forums, and manufacturers about the next generation wireless networks led to some common conclusions. According to them, 4G systems will become commercially available around 2010, when multimedia traffic and services will be dominant and prevail over voice services. Furthermore, an all-IP based network is foreseen, and this system will be entirely digital and packet-switched. The work toward 4G systems has already begun not only in academia but also in industry. The convergence of the different approaches into an efficient common system will be the measure of success.

References

[1]   http://www.etsi.org.
[2]   TIA/EIA-95-B, Telecommunications Industry Association (TIA) Standards.

[3]   Sollenberger, N. R., N. Seshadri, and R. Cox, AT&T Labs-Research "The Evolution of IS-136 TDMA for Third Generation Wireless Services," *IEEE Personal Communications Magazine,* Vol 6, June 1999, pp. 8–19.

[4]   http://www.itu.int/osg/imt-project/docs/What_is_IMT2000-2.pdf.

[5]   ITU-R Recommendation M.1455: "Key Characteristics for the International Mobile Telecommunications-2000 (IMT-2000) Radio Interfaces," May 2003.

[6]   Holma, H., and A. Toskala, *W-CDMA for UMTS (Revised Edition),* New York: John Wiley & Sons, 2001.

[7]   http://grouper.ieee.org/groups/802/11.

[8]   http://www.ofdm-forum.com/index.asp?ID=92.

[9]   http://www.etsi.org/frameset/home.htm?/technicalactiv/Hiperlan/hiperlan2.htm.

[10]   Paint, F., et al., "Mobility Aspects in 4G Networks—White Paper," Telenor R&D Report No. 43, December 2002.

[11]   Weiser, M., "Some Computer Science Issues in Ubiquitous Computing," *Communications of the ACM,* Vol. 36, No. 7, July 1993.

[12]   http://www.itu.int/home/imt.html.

[13]   http://www.itu.int/ITU-R/studygroups/rsg8/rwp8f/scope/index.html.

[14]   http://delson.org/4gmobile/intro.htm.

[15]   http://www.wireless-world-research.org.

[16]   WWRF "Book of Visions 2001—Version 1.0."

[17]   http://playground.sun.com/ipv6.

[18]   http://www.ipv6forum.com.

[19]   http://www.columbia.edu/itc/ee/e6951/2002spring/Projects/CVN/report2.pdf.

[20]   RFC 3314 "Recommendations for IPv6 in 3GPP Standards September 2002."

[21]   http://www.3gpp.org/About/about.htm.

[22]   http://www.3gpp2.org.

[23]   http://www.comsoc.org/~comt/msg00006.htm.

[24]   http://www.nttdocomo.com.

[25]   http://www.3gnewsroom.com/3g_news/oct_02/news_2563.shtml.

[26]   http://www.docomo.com.br/ingles/Vision.asp.

[27]   http://www.sun.com/index.xml.

[28]   http://research.sun.com/features/4g_wireless.

[29]   http://www.motorola.com/content/0,1037,276,00.html.

[30]   http://www.ericsson.com.

[31]   http://www.3gnewsroom.com/3g_news.

[32]   Gray, S. D., "Nokia's View on 4G," *International Forum on Future Mobile Telecommunications & China-EU Post Conference on B3G,* Beijing, China, November 20–22, 2002.

[33]   Wong, K., R. D. Murch, and K. B. Letaief, "Optimizing Time and Space MIMO Antenna System for Frequency Selective Fading Channels," *IEEE Journal on Selected Area in Communications,* Vol. 19, No. 7, July 2001, pp. 1395–1407.

[34]   CAUTION++ IST project (IST2001-38229) www.telecom.ece.ntua.gr/CautionPlus.

[35]   CAUTION++ IST Project, Deliverable "D-3.1: System Network Architecture," October 2003.

[36]   IST-MONASIDRE, IST project, http://www.monasidre.com.

[37]   IST-COMCAR, IST project, http://www.comcar.de.

[38]   Huschke, J., et al., "Dynamic Radio for Mobile Interactive Multimedia Systems," *European Wireless,* Dresden, Germany, September 12–14, 2000.

[39]   IST-DRIVE, IST project, http://www.ist-drive.org.

[40]   IST-OVERDRIVE, IST project, http://www.istoverdrive.org.

[41] IST-1999-10050 BRAIN Deliverable D2.2, "BRAIN Architecture Specifications and Models, BRAIN Functionality and Protocol Specification," 2001.

[42] IST-CAST, IST project, http://www.cast5.freeserve.co.uk.

[43] http://www.ece.gatech.edu/research/labs/bwn/projects.html.

[44] http://www.winlab.rutgers.edu/pub/docs/focus/MobNet2.html.

[45] Pereira, J., "Redefining Software (Defined) Radio: Reconfigurable Radio Systems and Networks," Special Issue on Software Defined Radio and Its Technologies, *IEICE Trans. Commun.*, Vol. E83-B, No. 6, June 2000.

[46] Beach, M. A., et al., "Reconfigurable Radio Systems and Networks," *IEE 3G Conference*, London, 2000.

[47] CAUTION IST-2001-38229, "D-2.2 System Requirements Specifications," May 2003.

[48] CAUTION IST-2001-38229, "D-3.1 System Architecture Definition," November 2003.

[49] Barbera, M., et al., "An Application of Case-Based Reasoning to the Adaptive Management of Wireless Networks," *Advances in Case-Based Reasoning*, Springer Verlag, Vol. 2416.

CHAPTER 6

Business Models for Wireless Systems Resource Management

6.1 The Challenge of Resource Management in Wireless Telecommunications Systems

This chapter focuses on the intimate relationships that exist between resource management and the optimization of wireless networks. The management of resources in wireless telecommunications systems is definitely one of the most complex optimization problems in current computer-based systems. The reasons for this complexity are to be found in the quick dynamics of the systems technological evolution and in the constant shifting of customers' expectations, in a market that accounts for huge potential revenues.

Resource management plays a very important role in wireless telecommunications systems and is in fact a continuous activity during the complete system's life cycle. When a wireless network is planned, each infrastructure investment is carefully dimensioned to match the resource requirements entailed by the offered traffic. As the network deployment proceeds, adjustments to the initial planning are made to tune the network with new data that become available from its real operation. During this period of network maturity, the optimization of resource usage becomes the key profit enabler for network operators, an activity supported by optimization tools whose license prices sometimes exceed that of the infrastructure itself.

The resource management activities described above are only referring to the topmost of the possible dimensions of the system's optimization. Another aspect that is an active research subject is the management of the radio resources at finer levels, which is mainly driven by the progressive enhancement of our ability to multiplex the communication flows over the available spectrum. Just to give an example of the significance of the enhancements that are being achieved, consider the increase in the bandwidth that has been made available in 2.5G systems with the introduction of EDGE technology over the existing GPRS systems.

In between these two types of resource management, there is another one, mainly related to the day-by-day administration of the wireless networks that refers to the ability of the system to dynamically cope with the variations in the offered traffic. Given a network architecture and a fixed transmission technology, even the small daily fluctuations in the traffic profile may have such an impact on the achievable profit that it could justify the deployment of more advanced and costly resource management schemes. In fact, in mature networks with millions of subscribers, the capability to rearrange the resource allocation for accommodating even 1% more of the offered traffic would result in significant increases in daily network revenues.

Meeting this challenge requires an investment that calls for a change of paradigm in the way networks are currently managed. It requires first of all, that the networks should be able to maintain a constant awareness of their operational environment, with improved monitoring capabilities that allow a real-time acquisition of information about service provision requests made by subscribers. Second, it entails an ability to dynamically and effectively react to predicted and unpredicted traffic variations, by applying fast and configurable resource allocation strategies that result in the most profitable exploitation of wireless network available resources.

The key concept we will be dealing with in this chapter is the profitability that can be achieved from dynamic resource allocation. Whereas other areas of network management, such as those mentioned above, have clear requirements that are easily translated into business results, the effectiveness of dynamic resource management has to be measured against a set of profitability criteria that should take into account the perspectives of various network stakeholders. More precisely, we deem of equal importance the perspective of the network operators and that of the network subscriber. The relevance of the first ones is made apparent from the role of service providers that network operators play in the wireless communication arena (though somebody may argue this is a biased situation that may change in the future), while that of the latter ones is achieving an increasing importance in current wireless telecommunications markets, for which customer retention is becoming a more and more challenging business objective.

The two aspects highlighted above are the basis for the definition of a set of mathematical models, which we call "business models," that we will propose in this chapter as a guide for the dynamic resource management of wireless networks. Business models are the tools we use to capture the various perspectives of the system stakeholders about the resource management choices of the system. Since such a set of models is a representation of the impact that the resource management choices have on the perceived system utility, they can be used early to assess the consequences of changes in resource assignment and definitely to drive the operations in a wireless telecommunications system.

The approach adopted to define the business models first identifies the relevant stakeholders of the system, and then the facets that each stakeholder would take into account if he or she were asked to assess the QoS provision. The various perspectives on the wireless network service provision from the different stakeholders are then combined into a mathematical structure that describes where value is generated in the economy of the wireless system.

The usefulness of the business models lies in the possibility of embedding into them the mathematical formulation of the resource management algorithms and thus to define optimization schemes for the network's operation. The input required for the business model definition imposes some requirements on the monitoring features of the wireless networks. In particular, network monitoring is requested to provide real-time detailed specifications of traffic profiles, which include estimates of the number of services and of the amount and type of resource requests. The solution of business models through proper algorithms will provide an operative support to the decision-making that wireless networks have to perform to be able to promptly and effectively react to changing traffic intensity and profile.

This chapter is organized as follows: We first introduce and discuss the perspectives of the main wireless network stakeholders, which should take into account in the definition of the business models. These perspectives are then combined into two different types of mathematical formulations, which are suitable to be applied at different levels of the decision-making process in resource management. The first type of formulations resorts to the concept of value structure, which is able to effectively handle high-level problems in resource assignment that involve multiple choices subject to multiple contrasting optimization criteria. The second type of mathematical formulation is based on a stochastic approach, which can adequately capture the intrinsic randomness of traffic patterns in wireless networks. Each of these two approaches is presented in a dedicated section, with detailed examples of applications. Moreover, we also discuss the operational deployment of the model into various wireless networks resource management schemes. Finally, a conclusion section is given, which also provides some experiences from the application of the proposed business models to real wireless networking environments.

6.2   Business Model Driving Criteria for Resource Management in Wireless Networks

This section discusses the criteria that should be taken into consideration when resource allocation decisions are to be made in wireless communications systems. A number of stakeholders exist in that arena, and the past, current, and future success of ubiquitous communications depends upon the possibility of simultaneously satisfying different interrelated needs, which include the ability to sustain the technological growth of communications networks, the ability to provide useful services through these networks, and least but not last the possibility to provide these services at an affordable price, with a friendly, simple, and unified access mechanism offered through a number of appropriate service access points.

The major stakeholders in the wireless telecommunications arena are those shown in Figure 6.1. All manufacturers, wireless network operators, providers, and subscribers, have their own perspectives in perceiving the costs and benefits of the wireless system, as well as their own business models with respect to the services they provide and receive, and each of them may be willing to "trade" some aspects more likely than other ones.

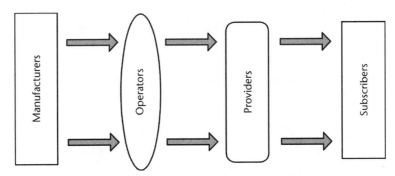

Figure 6.1   Wireless network stakeholders.

From a manufacturer's perspective, the value of the wireless telecommunications business is mainly affected by the market value of the offered product portfolio and by some key factors directly affecting production and maintenance costs, such as the technological evolution, reuse and deployment of either hardware, or software off-the-shelf components (COTS).

From a network operator and service provider perspective, the performance and dependability of the systems and services need to be taken into consideration to evaluate the amount of subscribers that can be satisfactorily served. These two metrics are indeed first translated into the two classical parameters defining the quality of a network service (i.e., call drop rate and call block rate) and ultimately in units of profit per deployed infrastructure element. Note that in the current wireless telecommunications world, network operators are in fact also covering the role of service providers.

Finally, the subscribers' perspective is mostly determined by a basic cost over quality ratio, which gets refined by a wide set of subtle parameters that range from aspects related to the type of service being requested, to the perceived subscriber friendliness of the service, until the hour of the day the request is made.

Since our purpose is that of analyzing and proposing models to support the dynamic resource management that should constantly take place in already deployed networks, we shall concentrate our attention on the two major network stakeholders in this scenario, namely the subscribers and the network operators. These two actors have quite different perspectives (which are, in most of the cases, ultimately contrasting) on the way the network should be managed. We will list the most relevant of these perspectives in Sections 6.2.1 and 6.2.2.

6.2.1 Subscribers' Perspectives

The perspective of the subscribers is definitely influenced by the limited, local view that the subscriber has (and quite fairly is entitled to have) regarding service provision. When in a call, or when transferring data through a wireless network, subscribers are satisfied if they can get good quality service themselves.

Therefore the fuzzy concept of quality becomes one fundamental factor for the definition of the subscriber's satisfaction. Definition of quality is strictly service-dependent. While the quality of a call is quite an obscure concept for GSM voice services, more precise definitions are needed to state the contract agreements for data services, which require peak and average throughput and packet loss probability quantifications.

From a high-level subscriber's perspective, some attributes of the service are commonly used to describe the perceived level of quality, such as service availability (readiness for usage), service reliability (continuous provision of a service), and performance (application layer observed throughput).

To simplify the subsequent treatment, it is possible to partition the level of service provision into a limited set of *modes*, where a mode is considered to be identified by a fixed amount of resources assigned for the service request fulfilment, and corresponds to a certain level of availability, reliability, and performance. This is actually a quite close representation of the way resource requests are handled in wireless networking management systems, where resource requests typically correspond to time slots in TDMA systems (GSM and GPRS) and to a limited set of possible connection speeds in CDMA-based systems (UMTS and WLAN).

Each service is associated with a set of limited modes through which it is provided. For instance, voice over GSM is a service that is currently being provided in two main modes (i.e., either with a full-rate connection or with a half-rate connection) with the subscriber's perceived quality in the first service provision mode being higher than the one in the second service provision mode. Real-time data services such as videoconferencing also have a few modes for service provision, whereas noninteractive data services such as telnet and e-mail may be provided in several modes while keeping the subscriber's satisfaction at quite acceptable levels.

An interesting aspect here is that the possible modes by which a service can be provided may be affected by the specific contract obligation of a subscriber. Indeed, there may be for instance premium subscribers for which, due to contract obligations, the service provision is possible only in an agreed mode.

In heterogeneous wireless networks, the perspective of the subscriber about service provision may be affected by additional factors. The network that provides the service may be one relevant factor, for the fact that the cost of the same service may vary in different networks. Moreover, there are classes of subscribers who are willing to pay more for increased quality, whereas there are classes of subscribers who will accept a lower quality for a less expensive service provision mode.

Also, it is worthwhile considering that the traffic load scenario may also affect the subscriber's perception of service provision. For instance, in an emergency situation the subscribers would be relatively much happier to receive a half-rate voice service than they would in normal situations. Moreover, the unavailability of videoconferencing services would most probably pass unnoticed in such a situation.

6.2.2 Network Operators' Perspectives

The views of the network operators are obviously influenced by the operational and maintenance costs of the networking infrastructure. The first and foremost wish of the network operators is to increase the revenues that can be obtained, and this can be accomplished through the maximization of the profit achieved per unit of the deployed network elements' capacity.

Notice that revenue maximization does not always pass through the maximization of resource utilization. This simplified approach indeed does not take into consideration that the cost for service provision that the subscribers are willing to pay is not a linear function of the allocated bandwidth, but it is instead a more complex function that depends on the quality that can be offered through that. For instance, it might be convenient in some scenarios to save some spare capacity of the network for the sake of ensuring a reliable and high-quality service provision that subscribers would be willing to pay for, rather than allocating the whole available bandwidth and then not being able to cope with unforeseen resource requests.

On the other hand, the ability of satisfying as much as possible of the offered traffic is one goal of network operators; in fact it entails in general a positive feedback effect that attracts further subscribers and improves customer retention. The attention of network operators' business strategy to customer's retention is gaining more and more importance as the wireless service market approaches a mature and stable state. Indeed, the end of the monopolistic market positions that existed till a few years ago has led to a very dynamic arena, in which the subscriber's fidelity is a dimming concept. Thus, the push for maximizing profit and resource utilization has to be tempered by a certain degree of fairness. This is even more important in certain traffic load scenarios, in which the value perceived by the subscriber for service provision is incommensurably higher than the one perceived by the network operator, and thus denying the service provision or selecting a poor provision mode may be counterproductive for the network operator in the medium to long range.

6.3 Value-Based Business Models

Since the decision-making related to resource management is in fact performed by the wireless network system itself, it makes sense to consider that business modeling is mainly inspired by the network operator's perspectives. However, as stated in the previous section, for a business model to be successful in the medium to long range, the subscriber's perspective must be taken into great consideration. On the other hand it is worth observing that this may not be needed for businesses that only consider a short-term time scale. For instance, a service provider's business model can be of a limited scope, since it is not bounded to huge network infrastructure investments such as those made by network operators.

Because of the profit-oriented perspective that our discussion will be focusing on, it makes sense to introduce at this stage the concept of *value,* as a unifying concept that is able to subsume the different perspectives of interest in this business modeling work. Value-based structures are used in decision-making approaches because of their generality and intuitiveness [1]. Moreover, they easily lend themselves to combine even quite different optimization facets into aggregate numerical structures that reflect the preferences of the various decisions, as it has been proven in several contexts [2–5]. Finally, automated tools exist to support the consistent definition of complex value structures [6].

The network operator's perspective on service provision can be easily transformed into a value. For each service request, the network operator has to know an estimation of the value achieved by satisfying the request. The value may depend on the mode that the network operator chooses to satisfy the request. For instance, let us consider again the GSM voice call service. If we let mode 1 be a full-rate channel for voice service provision and v_1 the value achieved, and mode 2 be a half-rate channel for voice service provision and v_2 the value achieved, we can easily expect that the relationship $v_1 \leq v_2$ is valid for the network operator. Obviously, the opposite relationship would be the one perceived by the corresponding part (i.e., the subscriber).

Let us summarize the various parameters that may affect the perspectives of both the subscriber and the network operator on service provision. They are the following:

1. The type of the service requested: Different expectations exist from a subscriber's perspective and different resource requirements may be entailed from a network operator's perspective, when varying the service.

2. The type of contract of the subscriber, which may impose constraints on the network operator.

3. The mode of service provision: The subscriber translates it into a quality level, whereas the network operator translates it into a resource assignment.

4. The radio access network that is used as a bearer for the service. This may impact either the cost or the quality of the service provision, thus affecting the value perceived by the subscriber. The perception of the network operator may differ from that of the subscriber.

5. The traffic load scenario, which, as we described above, may affect the relative importance that the service provision has for the subscriber and for the network operator.

These factors define several dimensions of the value structure. For the sake of simplicity, we shall limit these dimensions to four, by merging the first two, which together can be seen as defining the overall set of possible services (e.g., voice with guaranteed quality and videoconferencing available from 9 A.M. to 6 P.M. at 2 Mbps.

We shall now introduce some mathematical notations to make the subsequent treatment more rigorous. Thus, let i, $i = 1, 2, ..., n_S$ be the ith class of service the wireless network system is able to support, and let j, $j = 0, 1, 2, ..., n_M$ be the jth mode of service provision. Obviously, only a subset of the possible modes will be available for a given service type. However, we suppose that service mode 0, which corresponds to no service provision, is available for all service types. If the network selects service mode 0 for a service request, this means that it decides to deny the service. Moreover, let k, $k = 1, 2, ..., n_{RAN}$ be the kth type of radio access network that exists in the system and t, $t = 1, 2, ..., n_{TLS}$ the tth traffic load scenario the system can handle.

Thus, a full value structure for the resource management decision-making problem requires providing a value $v_{i,j,k,t}$ for each of the possible combinations of the four dimensions described above. As already mentioned, not all combinations are possible, as some service modes are only available for defined services, and also because there are services that can only be provided with a restricted set of radio access networks (e.g., highly demanding data services must go on UMTS or WLAN). Moreover, depending on which level the resource management takes place, some dimensions may not be relevant. For instance, when the decision making on how to allocate resources to voice and data calls is considered within a GSM/GPRS system, it is of no use to consider the impact on the value from the availability of other radio access networks.

Let n_i be the number of service provisions for which the resource management system has to make a decision, for service type i, $i = 1, 2, ..., n_S$. A service request can be satisfied in different modes, and thus a bandwidth request $c_{i,j}$ for the provision of a service request of type i, $i = 1, 2, ..., n_S$ in mode j, $j = 1, 2, ..., n_M$ must be known to the resource management entity. Moreover, since it may be in general possible to decide to satisfy a service provision over different radio access networks, the bandwidth request must also be known for each of the networks the system is able to manage. Thus, we shall add to the required bandwidth information a further dimension to take into consideration the dependency from the specific type of technology, and we shall denote with $c_{i,j,k}$ the bandwidth requested for the provision of a service request of type i, $i = 1, 2, ..., n_S$ in mode j, $j = 0, 1, 2, ..., n_M$ on radio access network k, $k = 1, 2, ..., n_{RAN}$.

Let $c_{i,j,k}$ denote the maximum bandwidth that can be made available for service type i, $i = 1, 2, ..., n_S$ in the network k, $k = 1, 2, ..., n_{RAN}$. The decision-making process has to respect the capacity constraints that reflect the maximum available bandwidth in each radio access network. Therefore, if we denote by $x_{i,j,k}$ the integer decision variable that represents the number of service provision requests of type i, $i = 1, 2, ..., n_S$ that the resource management system decides to serve with mode j, $j = 1, 2, ..., n_{RAN}$ on the radio access network k, $k = 1, 2, ..., n_{RAN}$ the following relationships must be satisfied for the decision represented by vector $\vec{x}$ in order to be a feasible solution to the optimization problem:

$$\sum_{i=1}^{n_S} \sum_{j=0}^{n_M} x_{i,j,k} \cdot c_{i,j,k} \le \sum_{i=1}^{n_S} Ci, k, k = 1, 2, ..., n_{RAN}$$

that is the total bandwidth requested from each radio access network cannot exceed the available one, and

$$\sum_{j=0}^{n_M} x_{i,j,k} \cdot c_{i,j,k} \le C_{i,k}, i = 1, 2, ..., n_S, k = 1, 2, ..., n_{RAN}$$

which states that the bandwidth assigned to each type of service request cannot exceed the bandwidth that may be allocated in a specific radio access network. Of course, the feasible values for variables $x_{i,j,k}$ are those and only those integers that fall within the interval $[0, n_j]$, $i = 1, 2, ..., n$, $j = 1, 2, ..., m$, $k = 1, 2, ..., n_{RAN}$.

The last constraint to be imposed on the optimization problem formulation is the one that requires the decision-making to produce a choice on each service request that the resource management receives, which is formulated as follows:

$$\sum_{j=0}^{n_S} \sum_{k=1}^{n_{RAN}} x_{i,j,k} = n_i, i = 1, 2, ..., n_S$$

The resource management entity should use the values that are adequate for the identified traffic load scenario t to instantiate its objective function $G_t(\vec{x})$ as follows:

$$G_t(\vec{x}) = \sum_{i=1}^{n_S} \sum_{j=0}^{n_M} \sum_{k=1}^{n_{RAN}} x_{i,j,k} \cdot v_{i,j,k,t}$$

The objective function and the constraints defined for the decision variables define together the mathematical formulation of the general resource management business model, which is given below in integer linear programming format.

$$
\begin{cases}
\text{Max } G_t(\vec{x}) = \sum_{i=1}^{n_S}\sum_{j=0}^{n_M}\sum_{k=1}^{n_{\text{RAN}}} x_{i,j,k} \cdot v_{i,j,k,t} \\[1em]
\text{subject to} \\[1em]
\sum_{j=0}^{n_M} x_{i,j,k} \cdot c_{i,j,k} \le C_{i,k}, \quad i=1,2,\ldots,n_S, k=1,2,\ldots,n_{\text{RAN}} \\[1em]
\sum_{i=1}^{n_S}\sum_{j=0}^{n_M} x_{i,j,k} \cdot c_{i,j,k} \le \sum_{i=1}^{n_S} Ci, k, k=1,2,\ldots,n_{\text{RAN}} \\[1em]
\sum_{j=0}^{n_S}\sum_{k=1}^{n_{\text{RAN}}} x_{i,j,k} = n_i, i=1,2,\ldots,n_S \\[1em]
x_{i,j,k} \in [0,n_j], xi,j,k \in N, i=1,2,\ldots,n_S, j=0,1,\ldots,n_M, k=1,2,\ldots,n_{\text{RAN}}
\end{cases}
$$

This general business model can be instantiated into different model subtypes, depending on the level at which the resource management takes place and on the amount of available information. In the following we shall consider an example of application for this business model, which is referring to a resource assignment decision-making problem inside the boundaries of a single radio access network (called the local resource management decision-making problem).

6.3.1   An Example of a Value-Based Business Model for the Radio Resource Management Decision-Making

In this section, the unit that performs the decision-making for the RRM, namely the resource management element as described in Section 5.3, can be considered to be an abstract node of the network management entity of a wireless system. This is connected to a monitoring subsystem that keeps it aware of the status of the network [7], in terms of the following:

- The values of a set of key performance indicators for the cell or access point in the congested area, which include traffic and signaling resources utilization;
- The number of service provision requests that are being made in the area;
- The types of service requests that are being made in the area, and the bandwidth requirement of each of them.

Obviously, the monitored parameters listed above may be, and in general are, dependent from the particular radio access network technology (GSM, UMTS, WLAN) that the resource management element is managing. To make the example more complete, we shall consider a GSM/GPRS system, in which both voice services over circuit connections and data services over packet connections are provided.

We will consider two types of services, namely voice calls and nonreal-time data transfers such as file download. Let us consider a typical resource assignment problem for a GSM/GPRS cell for which the resource management element may be asked to take a decision. Suppose that, at a given point in time, n_1 service requests are being made for voice calls and n_2 service requests exist for data services.

Each voice service request can be satisfied in two modes (i.e., with the allocation of a TCH with a full-rate transmission scheme or with the allocation of a half-rate TCH). In the first case the service provision has a cost for the system, in terms of resource consumption, of $c_{11} = 1$ measured in terms of GSM TCH channels, whereas for the second provision mode the cost is approximately $c_{11} = 1/2$.

Each data transmission service request can be served in 8 different modes, each mode corresponding to a number of allocated GPRS timeslots. Since each timeslot is corresponding to a GSM TCH, the cost in terms of bandwidth of the different service provision modes is $c_{2,j} = j, j = 1, 2, ..., 8$.

Let $C_1 = 30$ be the total bandwidth available in the cell for voice services (30 TCH traffic channels) and let $C_2 = 8$ (i.e., one carrier is assigned to GPRS service). The total available bandwidth of the cell is however such that C, $C_1 + C_2$, as some resources can be used for both GSM and GPRS services depending on the offered traffic. For our example, we assume that $C = 32$, which means that two of the TCH resources are reserved for GPRS traffic.

To specify a decision-making criterion that guides the resource management element in its selection, we shall use a simple value structure that assigns the following values:

- A normalized value $v_{11} = 1$ for the provision of a voice call over a full-rate TCH.

- A value $v_{12} = 0.8$ for the provision of a voice call over a half-rate TCH channel. Notice that the value is less than that achieved with a full-rate transmission, since it is expected that the subscriber's satisfaction will be negatively influenced by the reduced quality of the voice call.

- A value $v_{2,j} = j$ for the provision of a data transmission service with j timeslots, for $j = 1, 2, ..., 8$. This value is selected in such a way that considers that the satisfaction of the subscriber is affected by the speed of the transfer, which is proportional to the number of the allocated GPRS time slots. This is just an example, and more refined value assignments may be of course taken into consideration.

- A value $v_{1,0} = v_{2,0} = v_{j,0} = 0$ is given for service mode 0 (no service) for all requested services. Note that, in some scenarios, a choice $v_{1,0} = v_{2,0} = v_{j,0} < 0$ might be more appropriate to represent the fact that denying service provision adversely affects the system's utility.

The total value obtained by the system with the decision represented by vector $\vec{x}$ is given by function $G_t(\vec{x})$, which is defined as follows:

$$G_t(\vec{x}) = x_{1,1} + x_{1,2} \cdot 0.8 + \sum_{j=1}^{8} x_{2,j} \cdot j$$

The objective function and the constraints defined on the decision variables define together the mathematical formulation of the resource management element's business model, given below in integer linear programming format.

$$
\left\{
\begin{array}{l}
\text{Max } G_t(\vec{x}) = x_{1,1} + x_{1,2} \cdot 0.8 + \sum_{j=1}^{8} x_{2,j} \cdot j \\[2ex]
\text{subject to} \\[1ex]
\quad x_{1,1} + x_{1,2}/2 \leq 30 \\[2ex]
\quad \sum_{i=1}^{8} x_{2,j} \cdot j \leq 8 \\[2ex]
\quad x_{1,1} + x_{1,2}/2 + \sum_{j=1}^{8} x_{2,j} \cdot j \leq 32 \\[2ex]
\quad \sum_{j=0}^{8} x_{2,j} = n_2 \\[2ex]
x_{1,0}, x_{1,1}, x_{1,2} \in [0, n_1], \; x_{2,j} \in [0, n_2], \; j = 0, 1, \ldots, 8, \; x_{i,j} \in N, \forall i, j
\end{array}
\right.
$$

The optimization problem described above is an integer linear problem (ILP), whose feasible solutions, expressed by the values of variables $x_{i,j}$ define the set of decisions that the resource management element may take to solve the resource assignment problem. More precisely, any feasible solution to the resource management element's business model problem provides:

1. An assignment of resources for the service requests that are to be satisfied, which is defined by the nonzero values $x_{i,j}$, $j \neq 0$. These service requests will be those ones that the resource management element will try to accommodate in its controlled network.
2. A set of service provision requests that the resource management element will not serve, which are defined by the nonzero values $x_{i,j}$, $j = 0$.

The service requests at point (2) may be simply rejected, or they may be escalated to higher levels of the resource management decision-making. Indeed, even if the local decision-making process of the resource management element has decided that these services are not worth being served, it may be that additional resources are available at higher levels in the network management system. This might be the case for instance in heterogeneous wireless networks, where multiple radio access networks coexist and overlap the respective coverage areas. In this scenario, it makes sense to perform an additional attempt to accommodate the locally rejected requests into a different network.

The optimal solution for the ILP defined above is the one for which the objective function $G_t(\vec{x})$ returns the maximum value. This solution is also called the exact solution of the optimization problem. As we will describe in the following section, there exist algorithms for finding such an optimal solution.

6.3.2   Operational Deployment of Value-Based Business Models

Exactly solving the ILP relies on algorithms that extensively search the state of the feasible solutions by evaluating and comparing them in terms of the value taken by the objective function. A wide set of efficient approaches exist to limit such a search and to guide it towards the most promising solution, though it must be noticed that in the worst case the whole set of feasible solutions (which grows exponentially with the problem dimensions) must be searched.

Indeed, because the decision variables must only take integer values in the feasible solutions, the problem is in a form that classifies it as an NP-complete problem (i.e., a problem for which no polynomial time solution algorithm has been found so far [8]).

To explain the practical limitation that this theoretical result entails we will shortly describe one family of exact solution algorithms for ILP. Let us describe the space of feasible solutions as the leaves of a tree in which each level is a decision that assigns a value to a decision variable $x_{i,j}$, as shown in Figure 6.2. An exact solution algorithm for ILP performs a kind of a visit to the tree (the darkest path in Figure 6.2) without a criterion able to drive it directly towards the optimal solution. The only limiting criterion is the pre-evaluation of the partial solutions explored by the algorithm in its search. Indeed, under certain conditions it is possible to bound the value of each full solution that can exist in a given subtree. If the estimated value of the optimal solution in a subtree is lower than the current optimal solution explored by the search algorithm, then the whole subtree can be cut from the visit. This is the reason for which this kind of algorithm is often called a *branch-and-cut* algorithms.

However, it is important to remark that a number of approximate solution approaches are also possible [8]; these can return very good quality results without huge amounts of processing time. For instance, if some constraints on the decision variables are relaxed and more precisely when they are allowed to take any real value within the $[0, n_i]$ interval, then the problem becomes a much easier to solve, as it is a linear programming optimization problem, for which polynomial time solution algorithms exist.

In general, the particular algorithm to be applied depends on the form that the optimization problem assumes. As an example, it may be the case that the considered business model leads to a nonlinear definition of the objective function. This may be necessary to represent complex relative relationships between different parts

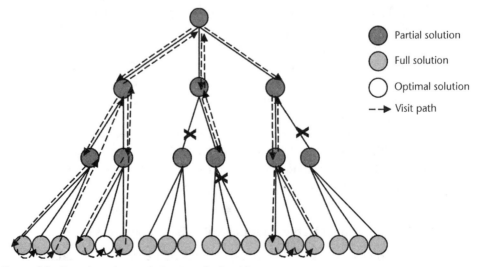

Figure 6.2 Branch-and-cut solution search algorithm.

of the overall decision—the case, for instance, when a given value is achieved only if two different services are simultaneously provided to the same subscriber with guaranteed levels of QoS. In this case, optimization algorithms other than those described here may be more conveniently applied.

6.4 Stochastic Business Models

A slightly different type of business model can be built to support other forms of resource assignment decision-making that must reflect the particular effects that the randomness of traffic patterns has on wireless networks. In this scenario, the purpose of the business model is still supporting the definition of the optimization criteria that guide the selection of the most suitable decision-making solution, but the evaluation of a given candidate solution is now performed through a stochastic model of the system.

The stochastic models we shall be considering in this section are those based on state-space representations of the wireless systems, such as Markov chain models and various types of Petri nets [9]. These models are very appropriate for representing the variability of traffic arrival patterns and of the resource requests made by each subscriber. Indeed, both the arrival of service provision requests and the time the resources are held are represented in these models through random variables, whose moments of various order can be appropriately set to reflect field observed data. Moreover, the logic of the resource assignment algorithms can be represented quite precisely thanks to the memory kept in the state space. The stochastic models make it possible therefore to reproduce, in a quite accurate fashion, the possible states of the modeled system and the transitions among those states.

In the specific case of wireless telecommunications systems, we will consider the following components for a representative stochastic business model:

- Traffic sources, which represent the arrival over time of service provision requests for the various classes of services that the system intends to provide;
- Subscriber requests, which are represented as active entities that move throughout the modeled system;
- Admission control logic for the system, which supervises the flow of request arrivals and denies access when resource availability does not allow the accommodation of any more requests (call blocking);
- Resource assignment logic of the system, which manages system resources and controls the evolution of subscriber requests that have passed admission control;
- The resources of the system, which are represented in a way to keep track of their availability status (free, allocated);
- Evaluation function, which determines the value of a resource management strategy in terms of some key network stochastic parameters, such as call block and call drop probability; usually defined as a performance metric of the modeled system [10].

The model elements are composed as shown in Figure 6.3, with the evaluation function able to obtain values from statistics about the network's model evolution.

The input data necessary for the instanciation of the stochastic business model consists of the following information:

- Network-observed conditions in terms of resource utilization at the time the decision-making is taking place. This data is to be obtained through network monitoring and is used to extrapolate the stochastic characterization of the traffic sources.
- Initial network configuration in terms of the maximum number of available resources in the modeled part of the network. This data is used to dimension the part of the resource management model.

The evaluation function may use values to weight the observed statistics so as to reflect the effect that the observed figures have on the network stakeholders. For example, the observed values of blocked and dropped call statistics do affect the subscriber's perception of the QoS, and so does the throughput of the connection. Conversely, the statistics that are most relevant for the network operator include those related to the rate of served requests, the average and peak utilization of the resource pools, and the throughput of the system in terms of service calls per time unit. Statistics and values may get combined through various types of mathematical functions. Usually, linear combinations are sufficient to capture the preferences to be represented into the business models.

6.4.1 An Example of the Stochastic Business Model for the Dynamic SDCCH Allocation Technique

The dynamic SDCCH resource management technique is a feature of GSM/GPRS networks that allows network operators to change the amount of traffic and signaling resources as seen in Section 2.5.3. In the dynamic SDCCH allocation, the mapping between logical channels and physical channels in a single cell is varied so that the amount of channels used for signaling (SDCCHs) is increased at the expenses of TCHs. Each time slot in a TCH can be reallocated for the purposes of signaling and be used as an SDCCH. The problem of devising the optimal number of additional SDCCHs obviously depends on the amount of traffic T_i (expressed for instance in Erlangs) at the specific cell, and must take into consideration the balance between

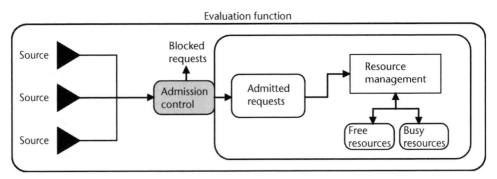

Figure 6.3   Structure of a general stochastic business model.

the decrease in the SDCCH utilization and the increase in the utilization of TCHs. Both these factors affect the network stakeholders, as we will explain in the following.

The dynamic SDCCH resource management technique has two principal objectives:

1. Reduce network congestion on SDCCH resources, without inducing it on TCH ones;
2. Enhance the capacity to guarantee the subscriber's access to the network.

To reduce network congestion means to decrease the SDCCH utilization factor, whereas, to enhance the capacity for accessing the network means increasing the availability of resources for a wider population of subscribers thus achieving a better subscriber's satisfaction. Increasing the number of subscribers that successfully access the network implies a growth of service rate and a reduction of blocking probability leading to higher operator profits as well. Because of its relevant effectiveness in satisfying network operators' and subscribers' wishes simultaneously, this resource management technique can be applied in every traffic load scenario that exhibits congestion on signaling resources, when a residual availability of traffic resources still exists.

A business model for the application of this technique may be employed when the unit has already made a decision that a part of the traffic is to be sustained by the increased capacity effect of the dynamic SDCCH resource management technique. The business models support the decision regarding which is the optimal number of channels to be transformed from TCH to SDCCH resources, a process that is called *fine-tuning* of the resource management technique [11].

Let us suppose that the dynamic SDCCH allocation technique has been selected as a solution for alleviating the congestion in a set of congested cells $C = \{1, ..., n\}$. Since any modification in the assignment of logical channels can be decided on a cell basis, we can separate the problem and perform a local optimization.

The effectiveness of the dynamic SDCCH resource management technique is measured in terms of growth on the service rate and the decrease on the blocking probability, which both lead to an improved subscriber's satisfaction and operator's profits. This leads to the definition of a business model that is driven by the following optimization function G:

$$G = w \cdot Oper\_gain + (1 - w) \cdot User\_sat$$

which is defined as a weighted linear combination of network operator's profit and subscriber's satisfaction contributions. The weight w should be selected so that the term to be privileged is actually the most important for the situation at hand. A dependency of w on the identified traffic load scenario appears as a sensible choice. The gain of the network operator is defined in terms of measurable network parameters as follows:

$$Oper\_gain = TCH_{th} \cdot E[call\_duration] + SDCCH_{SMS\_th}$$

where TCH_{th} is the throughput of the GSM cell system, observed at the TCH level (average number of TCH resource requests satisfied per unit of time), $E[call\_duration]$ is the average value for the duration of a GSM call, and $SDCCH_{SMS\_th}$ represents the part of the SDCCH throughput that is devoted to the SMS service (average number of SMS requests satisfied per unit of time). The subscriber's satisfaction represents the grade of satisfaction experienced by the subscribers and it is expressed by a percentage as follows:

$$User\_ sat = 1 - P_{Block\_TCH} - P_{Block\_SDCCH}$$

where P_{Block_TCH} and P_{Block_SDCCH} represent the blocking probabilities on TCHs and SDCCHs channels, respectively.

The business model defined by function G guides the selection of the optimal number of additional SDCCH channels (and the consequent reduction of the TCH ones) that best serve the traffic situation. To evaluate a possible solution to the problem, the Petri net model shown in Figure 6.4 can be used to estimate the proper allocation of channels. The traffic T_i at the cell i can be converted into an equivalent arrival rate, which is assigned to transition arrival. Initiating the arrival's transition models the request for a subscriber to establish a call, and generates a token that is put in the output place. At this point, the call can be either granted an SDCCH or blocked, depending on the availability of SDCCH and TCH channels. Blocked calls are lost. Calls that get one SDCCH conclude the set up and reach the TCH granting phase. After the call's termination, the TCH is released.

Given a description of the offered traffic profile, the Petri net model can be instanciated and solved quite easily. The solution of the model provides the statistics that will be used by the evaluation function to assess the value of the candidate solution.

In the particular case of this resource management technique, the candidate solution is a proposed allocation of dynamic SDCCH channels. The Petri net solution returns the statistics of the cell that would be observed if the proposed solution was actuated on the real network. The evaluation function will then return the overall value of the candidate solution, which can be used in the operational problem solution scheme proposed in Section 6.4.2.

6.4.2   Operational Deployment of Stochastic Business Models

Solving the optimization problem defined by the business model requires to repeatedly solving the Petri net model to explore the state of the feasible solutions.

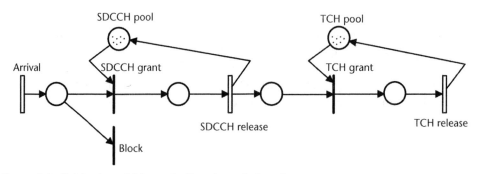

Figure 6.4   Petri net model for evaluating channel allocation.

Similarly, for the value-based business models, there does not exist a criterion able to guide such an exploration of the state of possible solutions. However, it has been experimentally observed that, for most of the considered business models evaluation functions, the problem shows some concavity properties, which means that it usually exhibits a single maximum point (or minimum, depending on the form of the evaluation function), which makes it possible to quickly find optimal solutions with gradient-based iterative algorithms.

The computation of the Petri net model basically requires solving a system of linear equations, whose number is a function of some model parameters, typically of the number of the logical channels with which a cell is equipped. For the example given in this section, if n_{TCH} and n_{SDCCH} are the number of TCH and SDCCH channels in a cell, computing the statistics on the cell requires solving a linear system of about $m = n_{\mathrm{TCH}} \cdot n_{\mathrm{SDCCH}}$ linear equations, at a computational cost of about [in terms of the costliest operations from a computational viewpoint (i.e., multiplications)]. The gradient solution schemes usually require a few steps to achieve quite a satisfactory approximation of the optimal solution, but it must be remarked that the concavity/convexity of the problem is not guaranteed.

Depending on the allowable time for the decision-making, this computational cost may be excessive and not feasible for a run-time solution. However, since the number of possible solutions for this particular optimization problem is comparatively small (remember we are dealing with just one cell at a time because of our assumption for a local impact resource management techinque), the Petri net model can be solved off-line with a tool by varying its parameters to maximize the optimization function defining the business model. The obtained results can be conveniently stored in a lookup table and used at run time to retrieve the optimal result for the SDCCH allocation for the current traffic situation at the cell.

This last approach has been successfully experienced in the context of the IST project CAUTION [12], in which various stochastic business models have been developed and used at run-time to drive some steps of the decision-making for resource management.

6.5 Conclusions

This chapter deals with the criteria that should be posed as the basis of any resource management decision-making process in wireless networking systems. In particular, we have focused on a particular kind of dynamic resource management that is currently not performed on a regular basis in wireless networks, but that has a huge potential from a revenue point of view.

The approach presented here refers to the ability of the system to dynamically cope with the variations in the offered traffic. Traffic fluctuations occur constantly in wireless networks. Each day, the traffic rises in a predictable manner during the busy hours and then diminishes to reach its minimum level during the late night hours. Moreover, the variation pattern changes along the days of the week, and longer variations cycles can be identified over a year's time. Finally, spikes of service provision requests may occur as a result of particular events that tend to concentrate subscribers in small areas or from sudden increases in the arrival rates of calls (e.g., emergency situations).

These variations in the traffic profile are hardly handled by the current resource management systems of wireless networks, and as a result, a part of the offered traffic is not served. To properly manage the exceeding traffic, an unattended network may deploy a means of automated resource management, which gets triggered when increases in the service provision requests occur, which, in turn, may actuate a proper reaction to effectively alleviate congestion effects while enhancing the network's ability to satisfy the subscribers' requests.

Towards this goal, the wireless network's resource management system needs to be endowed with automated tools that can drive the decision-making process properly and actuate the most effective resource management technique. The business models presented in this chapter have the purpose of fulfilling this need.

We presented different types of business models, which are suitable to support the network's optimization problem at various levels and in various scenarios. Value-based business models are particularly useful when multiple contrasting criteria are to be taken into consideration; the value is indeed a powerful unifying concept that can support efficiently the definition of overall optimization objectives. Stochastic business models are more adequate to lower level decisions, in which some more refined information concerning typical network key performance indicators, such as blocking probability, drop probability, and average service rate must be taken into consideration.

The first kind of business model has been proposed in the context of wireless network systems that comprise heterogeneous radio access network technologies [13]. In the framework of the IST project CAUTION++ [14], the resource management problem is stated in a very general context, with all the resources of GSM/GPRS, UMTS, and WLAN systems participating in a global network optimization. In this context, the general applicability of value structures is pointed out. Value structures play an important role in the definition of a uniform optimization strategy that fits the needs of this heterogeneous and diversified environment.

The latter type of business model has been already proposed and applied in the IST project CAUTION [12], where it was effectively demonstrated for the fine-tuning of the resource management techniques that were available to the resource management unit of the system. A number of business models were developed therein, one for each resource management unit. In some cases, business models were developed even for resource management techniques that did not require a fine-tuning procedure (i.e., they had no parameters to be set), for the purposes of better understanding the impact of their application on the network and defining the limits of the resource management technique's applicability in terms of the optimal achievable gains.

The business modeling activities reported in this chapter may be extended along different directions, and other types of business models could be considered as well. Among the interesting types of business models that could be of interest we can mention those based on neural networks. Indeed, neural networks are increasingly being used in real-world business applications, and, in some cases they have already become the method of choice (e.g., for fraud detection) [14]. Another type of a business model that could effectively support the type of decision-making required for the resource management of interest in our scenario is based on the Markov decision process [15]. However the modeling tools that we examined in detail here are considered to be of more direct application, also because the target market of the resource management tools (i.e., network operators) is more ready to accept well-consolidated optimization tools such as those based on classical operations research and stochastic models.

References

[1]   Suppes P., et al., *Foundations of Measurement: Geometrical, Threshold and Probabilistic Representations*, New York: Academic Press, 1992.

[2]   Vincke P., *Multicriteria Decision-Aid*, New York: John Wiley and Sons, 1992.

[3]   Bestavros, A., and S. Nagy, "Value-Cognizant Admission Control for RTDB Systems," *IEEE Real Time Systems Symposium*, Washington, D.C., 1996, pp. 230–239.

[4]   Bondavalli A., F. Di Giandomenico, and I. Mura, "An Optimal Value-Based Admission Policy and its Reflective Use in Real-Time Systems," *Real-Time Systems Journal*, January 1999.

[5]   Di Giandomenico, F., et al., "Scheduling Solutions for a Unified Approach to the Tolerance of Value and Timing Faults," in *Proceedings of Fast Abstract Session at the Twenty-Ninth Fault-Tolerant Computing Symposium, FTCS'99*, Madison, Wisconsin, June 1999.

[6]   Saaty T. L., *Multiple Criteria Decision Making: The Analytic Hierarchy Process*, RWS Publications, 1992.

[7]   CAUTION++ Deliverable D3.2, "Specification of Traffic Monitoring Components," CAUTION++ consortium, 2003.

[8]   Papadimitriou, C. H., *Combinatorial Optimization: Algorithms and Complexity*, Upper Saddle River, NJ: Prentice-Hall, 1982.

[9]   Ajmone-Marsan, M., G. Conte, and G. Balbo, "A Class of Generalized Stochastic Petri Nets for the Performance Evaluation of Multiprocessor Systems," *ACM Transactions on Computer Systems*, Vol. 2, 1984, pp. 93–122.

[10] Meyer, J. F., "On Evaluating the Performability of Degradable Computer Systems," *IEEE Transactions on Computers*, Vol. C-29, 1980, pp. 720–731.

[11] CAUTION++ Deliverable D3.4, "Decision Making Process Specifications," the CAUTION++ consortium, 2003.

[12] CAUTION Deliverable D4.2, "System Implementation Report," the CAUTION consortium, 2002.

[13] CAUTION++ Deliverable D3.1, "System Architecture Definition," the CAUTION++ consortium, 2003.

[14] Vellido, A., and P. J. G. Lisboa, *Business Applications of Neural Networks*, Liverpool, United Kingdom: John Moores University & Bill Edisbury Power Limited, 2000.

[15] Puterman M. L., *Markov Decision Processes*, New York: John Wiley and Sons, 1994.

CHAPTER 7

Conclusions

This book has demonstrated the importance of radio resource management and its application in wireless systems of present and future generation networks. In-depth studies in operational cellular systems were described, focusing on the limitations of current networks and the possible ways for alleviating their shortcomings. A projection of these limitations to next generation wireless systems shows the constraints of future networks as well.

Chapter 2, the first technical chapter, provided an overview of 2G networks. This was followed by a thorough network performance evaluation that was based on real measurements, both in normal and congested situations. The evaluations highlighted the limitations of current networks and focused on what is triggering traffic congestion. More specifically, it was shown that despite the heavy congestion that is experienced frequently, cellular systems are most of the times not fully utilized, and there exist many cases that can be characterized as very underutilized. The fact that these systems cannot be fully exploited, despite the efforts from standardization bodies and the research and industrial community, allows for additional research studies that aim to develop innovative radio resource management systems and techniques that can handle the problem more efficiently. Such an approach is independent of the wireless network, and it focuses on the real-time monitoring of characteristic KPIs, the alarming mechanism, and the intelligent and adaptive RRM system that executes the techniques as long as the congestion is present. In Chapter 2, the major KPIs for 2G networks were also presented and stability thresholds were discussed. It was observed that the KPIs start to show instability, when the values exceed a specific threshold. When these thresholds are reached, a KPI indicates congestion and a chain-effect is observed in a way that worsens in one way or another all KPIs. For that purpose, a number of traffic load scenarios was identified, each one describing a congestion situation. Furthermore, several resource management techniques were presented. Some of these techniques are considered to be innovative, while others make use of standardized features that can be exploited, if they are adjusted at a real-time basis.

The experience gained by our work on 2G systems was used as a basis for studying the 2+G characteristics and proposing respective RRM techniques for applying efficient resource management. One of the major observations so far, is that the performance of 2+G systems depends on the performance of the circuit-switched part, mainly voice services, which allows the further development of RRMs that focus on the circuit-switched part to enhance the packet-switched one. A new characteristic in packet-switched wireless systems is the effect of system management on the fixed part that is mainly IP-based, in the overall performance. Chapter 3 presented the main characteristics of 2+G systems, focusing on the GRPS system. Planning characteristics of GPRS were presented, since a good overview is required to understand the effect of congestion in this kind of network. The real-time monitoring prerequisites for GPRS were also discussed, by presenting the characteristic KPIs. Finally, resource management techniques, focusing both on the air interface and the fixed part were analyzed.

The evolution of wireless communications and the need for a universal system led us to 3G. The implementation and deployment of 3G was and continues to be a difficult milestone for the wireless community, and this is strongly linked with the crisis that is experienced in the area of telecommunications during the first years of the new century. 3G systems aim to fulfill the advanced user requirements for multimedia high-speed communication. Chapter 4 presented standardized releases of 3G along with their characteristics. In addition, Chapter 4 described the task of 3G planning described, since it is an important step toward the effective deployment of the system. The purpose of describing the planning mechanism is to show the major features of the technology that is utilized for the radio access network of 3G, namely CDMA. In 3G networks, radio resource management is considered to be an important entity of the system. Moreover, RRM techniques were introduced and described in terms of their effectiveness.

Despite the fact that the deployment of 3G systems has not reached the desired level, the research and industrial community is searching for new solutions to bridge the gaps between the various wireless networking technologies, addressing multiple access network technologies. Therefore, we talk about systems beyond 3G and 4G. The definition of these terms is rather difficult, since most of the manufacturers and the operators have different definitions. Chapter 5 described the trends toward systems beyond 3G and the major research activities in these areas. As mentioned in the introduction, the means to achieve increased network performance by decreasing the effect of traffic overload is to enable a real-time monitoring system and an intelligent decision-making component, transparent of the resource management techniques, where these can be implemented. This system concept is described as a strong solution toward global RRM. This system is able to monitor a number of access networks and utilize the operator's resources in a way to guarantee both increased income and user's satisfaction.

Chapter 6 presented business modeling for the purpose of radio resource management. As described in the previous chapters, several techniques aim to solve the congestion problems, even if sometimes these are linked with a degradation of the offered service or the operator's gain. For that purpose, business models should be considered before the execution of a management technique. Each technique can be considered as the adjustment of a number of parameters. Each adjustment has a direct influence on the effectiveness of the system. Therefore, the trade-off between satisfaction and gain should be described in a scalable business model. Moreover, the suggestion of such a hierarchical RRM system is to hand over some traffic from one access network to another, while the definition of the percentage is a difficult task that should consider various aspects. Additionally, such a decision might lead to the shift of traffic from one operator to another, if the other operator has adequate resources and the user can afford this, while the process should be transparent.

In conclusion, we strongly believe that this book provides useful material for wireless networking researchers and operators, since it is based on measurements and experimentations in real networking environments that prove its importance.

List of Acronyms

3GPP Third Generation Partnership Project

ACCH associated control channel

ALCAP access link control application part

ATM asynchronous transfer mode

AuC authentication center

BCCH broadcast control channel

BH busy hour

BLER block error rate

BMC broadcast/multicast control

BR blocking rate

BSC base station controller

BSIC base station identification code

BSS base station subsystem

BTS base transceiver station

CAMEL customized application for mobile network enhanced logic

CCC common coordination channel

CCCH common control channel

CCITT Comité Consultatif International Téléphonique et Télégraphique

CDMA code division multiple access

CEPT Conférence Européene des Postes et des Télécommunications

CIR committed information rate

CM connection management

CN core network

CS coding scheme

CS circuit-switched

CSSR call set up success rate

CTS cordless telephony system

DAB digital audio broadcast

DAMPS digital advanced mobile phone service

DCH dedicated transport channel

DCR drop call rate

DL downlink

DVB digital video broadcast

DW   network data warehouse

EDGE   enhanced data rates for GSM evolution

EFR   enhanced full rate

EGPRS   enhanced general packet radio service

EIR   equipment identity register

eMLPP   enhanced multilevel precedence and preemption

ETSI   European Telecommunications Standards Institute

FCCH   frequency correction channel

FCFS   first-come first-served

FDD   frequency division duplex

FDMA   frequency division multiple access

FM   fault management

FTP   file transfer protocol

GERAN   GSM EDGE radio access network

GGSN   gateway GPRS support node

GMSC   gateway MSC

GMSK   Gaussian minimum-shift keying

GMU   global management unit

GOS   grade of service

GPRS   general packet radio service

GSM   global system for mobile communications

GSN   GPRS support node

HHO   hard handover

HLR   home location register

HO   handover

HOSR   handover success rate

HSCSD   high-speed circuit-switched data

IMEI   international mobile equipment identity

IN   intelligent network

IP   Internet protocol

IS-95   interim standard '95

ISDN   integrated services digital network

ISO   International Standards Organization

ITMU   interface traffic monitoring unit

ITU   International Telecommunications Union

KPI   key performance indicator

LA   location area

LBFS   least bits left first served

LLC   logical link control

LNA   low noise amplifier

MAC   medium access protocol

MAP   mobile application part

MAT   modified advanced time

MED   modified earliest deadline

MExE   mobile execution environment

MIMO   multiple input multiple output

MLT   minimum laxity threshold

MM   mobility management

MOC   mobile-originated calls

MS   mobile station

MSC   mobile switching center

MSRN   mobile station roaming number

MTC   mobile-terminated calls

NBAP   node B application part

NCH   notification channel

NMS   network management system

NMT   Nordic Mobile Telephone

OMC   operation and maintenance center

OSI   open systems interconnection

OS   operation subsystem

OVSF   orthogonal variable spreading factor

PBCCH   packet broadcast control channel

PCCCH   packet common control channel

PCH   paging channel

PCU   packet control unit

PDA   personal digital assistant

PDCH   packet data channel

PDCP   packet data convergence protocol

PDN   packet data network

PDTCH   packet data traffic channel

PDU   protocol data unit

PLMN   public land mobile network

PM   performance measurement

PS   packet-switched

PSK   phase shift keying

PSTN   public switched telephone network

QAM   quadrature amplitude modulation

QoS   quality of service

RACH   random access channel

RAN   radio access network

RANAP   radio access network application part

RAT   radio access technology

RLC   radio link control

RMU   resource management unit

RNC   radio network controller

RNS   radio network subsystem

RR   radio resource

RRM   radio resource management

SA   services and system aspects

SACCH   slow associated control channel

SAP   service access point

SCH   synchronization channel

SDCCH   stand-alone dedicated control channel

SDR   software-defined radio

SGSN   serving GPRS support node

SHO   soft handover

SMG   Special Mobile Group

SMS   short message service

SNDCP   subnetwork dependent convergence protocol

SRNS   serving radio network subsystem

SS7   signaling system 7

SSDT   site selection diversity transmission

TBF   temporary block flow

TC   traffic control

TCH   traffic channel

TCH/FR   full-rate TCH

TCH/HR   half-rate TCH

TDD   time division duplex

TDMA   time division multiple access

TFI   temporary flow identities

TNCP   transport network control plane

TNUP   transport network user plane

TR   technical report

TRAU   transcoding and rate adaptation unit

TRX   transceiver

TS   technical specification

TSG   technical specification groups

TSL   time slot

UE   user equipment

UL   uplink

UMTS   universal mobile telecommunications system

UP   user plane

UTRA   UMTS terrestrial radio access

VHE   virtual home environment
VLR   visitor location register
WAP   wireless application protocol
WCDMA   wideband CDMA
WLAN   wireless LAN
WPA   wideband power amplifiers
WWRF   Wireless World Research Forum
WWW   World Wide Web

About the Authors

Sofoklis A. Kyriazakos was born in Athens, Greece, in 1975. He studied electrical engineering at RWTH in Aachen, Germany, and specialized in wireless systems at the chair of Communications Networks at the same university. He then moved to the Telecommunications Laboratory of the National Technical University of Athens (NTUA), Greece, where he received his Ph.D. in the area of resource management in wireless networks. He is currently a senior research associate in the Telecommunications Laboratory of NTUA and participates in various research projects in the area of wireless telecommunications. He has served as a member of technical committees and as a reviewer for various conferences, and has authored book chapters as well as many publications in journals, conferences, and standardization bodies.

George T. Karetsos was born in Karditsa, Greece, in 1968. He obtained a degree in electrical and computer engineering in 1992 and a Ph.D. in telecommunication systems in 1996, both from the NTUA. He is an assistant professor in the Information Technology and Telecommunications Department of the Technological Education Institute of Larissa, Greece, and a senior research engineer in the Telecommunications Laboratory of NTUA. He has participated in many European and national research projects dealing with the optimization of protocols and operations as well as with the efficient design and management of advanced services in fixed and wireless networks. He has served as a member of technical committees and as a reviewer of various international conferences and journals. His research interests are in the areas of active networking, nomadic computing, performance evaluation, and resource management for fixed and wireless networks. He has published more than 60 papers in journals and conferences.

Index

For further information on these and other Artech House titles, including previously considered out-of-print books now available through our In-Print-Forever® (IPF®) program, contact:

Artech House
685 Canton Street
Norwood, MA 02062
Phone: 781-769-9750
Fax: 781-769-6334
e-mail: artech@artechhouse.com

Artech House
46 Gillingham Street
London SW1V 1AH UK
Phone: +44 (0)20 7596-8750
Fax: +44 (0)20 7630-0166
e-mail: artech-uk@artechhouse.com

Find us on the World Wide Web at:
www.artechhouse.com